About this book

About your BTEC First Applied Science

Choosing to study for a BTEC First Applied Science qualification is a great decision to make for lots of reasons. More and more employers are looking for well-qualified people to work within the fields of science, technology, engineering and maths. The applied sciences offer a wide variety of careers, such as forensic scientist, drug researcher, medical physics technician, science technician and many more. Your BTEC will sharpen your skills for employment or further study.

Your BTEC First Applied Science is a vocational or work-related qualification. This doesn't mean that it will give you all the skills you need to do a job, but it does mean that you'll have the opportunity to gain specific knowledge, understanding and skills that are relevant if you go onto further study or into technician-related employment.

What will you be doing?

Principles of Applied Science is the first of the two BTEC Applied Science Awards. The second is Application of Science. In both Awards you will be covering all aspects of science, including biology, chemistry and physics, as well as maths and health and safety-related issues. You will complete assignments based on scientific job-related scenarios, for example, working as a science technician, producing information for the public and conducting scientific research. As well as exploring a range of scientific concepts, you will use your IT skills to produce documents for assessment. Other skills you will practise include researching, preparing and giving presentations, producing scientific reports, following instructions for practical investigations and effective time management.

About the authors

David Goodfellow is a freelance writer and examiner, having previously taught chemistry at all levels for over 20 years. He led the development of the AS Science qualification for 2007 and is an experienced author.

Sue Hocking has been an examiner for almost 30 years. She has delivered BTEC science and health studies courses in FE colleges, as well as GCSE and A level biology courses in secondary schools and a sixth form college. Her specialist fields are biology, biochemistry and health promotion. Sue was the series editor for OCR A level Biology and has written many books and teacher support resources.

Dr Ismail Musa is a standards verifier for BTEC Applied Science (Level 2 and 3), an examiner for both GCSE and A level Physics and was involved in the development of the 2010 specification for BTEC Applied Science. Ismail has been teaching vocational applied science courses for over 12 years. As well as teaching and examining he has been working as a Subject Learning Coach for Science (SLC), coaching students and staff and organising and delivering teaching and learning sessions.

Contents

BTEC FIRST AWARD

PRINCIPLES OF APPLIED SCIENCE

David Goodfellow • Sue Hocking • Ismail Musa

ALWAYS LEARNING

PEARSON

Published by Pearson Education Limited, Edinburgh Gate, Harlow, Essex, CM20 2JE.

www.pearsonschoolsandfecolleges.co.uk

Copies of official specifications for all Edexcel qualifications may be found on the Edexcel website: www.edexcel.com

Edited by Ashwell Enterprises Limited, Nancy Hillelson, Priscilla Goldby and Tim Jackson
Designed by Wooden Ark Limited and Andy Magee
Typeset by TechSet

Illustrated by TechSet
Cover design by Pearson Education Limited
Picture research by Rebecca Sodergen

First published 2012

16 15 14
10 9 8 7 6 5 4

British Library Cataloguing in Publication Data
A catalogue record for this book is available from the British Library

ISBN 978 1 446 90279 0

Printed in Slovakia by Neografia

Acknowledgements
We would like to thank Hannah Verghese, Advocacy and Policy Manager, for permission to include information on The Migraine Trust.

A note from the publisher
In order to ensure that this resource offers high-quality support for the associated BTEC qualification, it has been through a review process by the awarding organisation to confirm that it fully covers the teaching and learning content of the specification or part of a specification at which it is aimed, and demonstrates an appropriate balance between the development of subject skills, knowledge and understanding, in addition to preparation for assessment.

While the publishers have made every attempt to ensure that advice on the qualification and its assessment is accurate, the official specification and associated assessment guidance materials are the only authoritative source of information and should always be referred to for definitive guidance.

BTEC examiners have not contributed to any sections in this resource relevant to examination papers for which they have responsibility.

No material from an endorsed student book will be used verbatim in any assessment set by BTEC.

Endorsement of a student book does not mean that the student book is required to achieve this BTEC qualification, nor does it mean that it is the only suitable material available to support the qualification, and any resource lists produced by the awarding organisation shall include this and other appropriate resources.

Unit 4 Biology and Our Environment

How to use this book

This book is designed to help you through your BTEC First Applied Science Award in Principles of Applied Science. It is divided into four units to reflect the units in the specification. This book contains many features that will help you use your skills and knowledge in work-related situations and assist you in getting the most from your course.

Introduction

These introductions give you a snapshot of what to expect from each unit – and what you should be aiming for by the time you finish it.

Learning aims

Learning aims describe what you will be doing in this unit.

Learner voice

A learner shares how working through the unit has helped them.

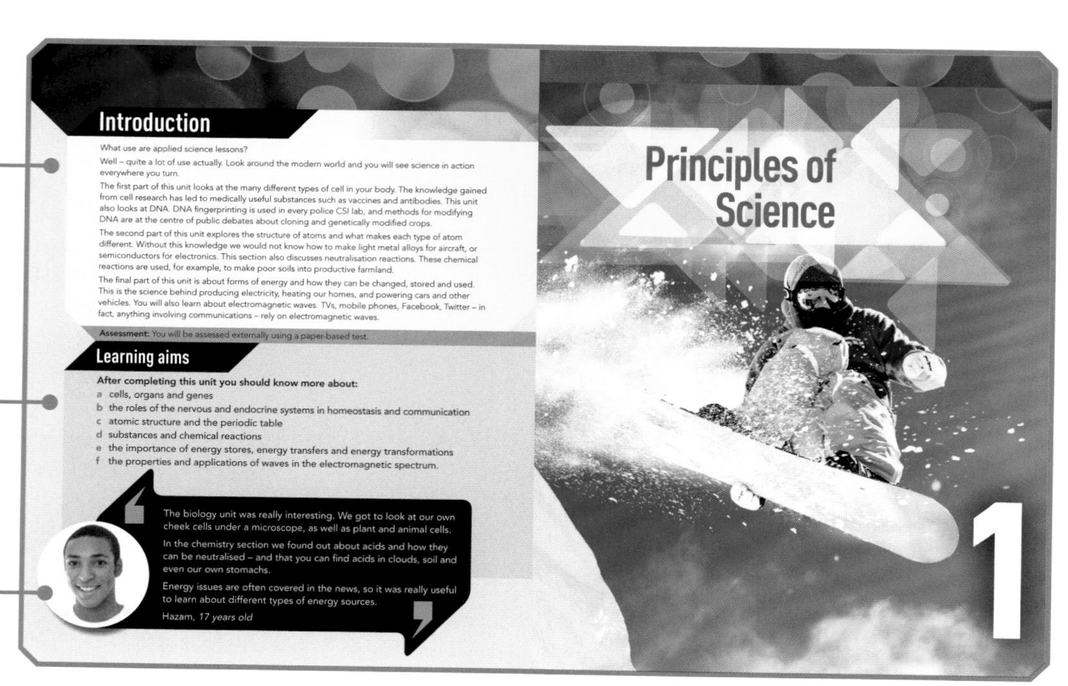

Introduction

What use are applied science lessons?

Well – quite a lot of use actually. Look around the modern world and you will see science in action everywhere you turn.

The first part of this unit looks at the many different types of cell in your body. The knowledge gained from cell research has led to medically useful substances such as vaccines and antibodies. This unit also looks at DNA. DNA fingerprinting is used in every police CSI lab, and methods for modifying DNA are at the centre of public debates about cloning and genetically modified crops.

The second part of this unit explores the structure of atoms and what makes each type of atom different. Without this knowledge we would not know how to make light metal alloys for aircraft, or semiconductors for electronics. This section also discusses neutralisation reactions. These chemical reactions are used, for example, to make poor soils into productive farmland.

The final part of this unit is about forms of energy and how they can be changed, stored and used. This is the science behind producing electricity, heating our homes, and powering cars and other vehicles. You will also learn about electromagnetic waves. TVs, mobile phones, Facebook, Twitter – in fact, anything involving communications – rely on electromagnetic waves.

Assessment: You will be assessed externally using a paper-based test.

Learning aims

After completing this unit you should know more about:

a cells, organs and genes
b the roles of the nervous and endocrine systems in homeostasis and communication
c atomic structure and the periodic table
d substances and chemical reactions
e the importance of energy stores, energy transfers and energy transformations
f the properties and applications of waves in the electromagnetic spectrum.

The biology unit was really interesting. We got to look at our own cheek cells under a microscope, as well as plant and animal cells.

In the chemistry section we found out about acids and how they can be neutralised – and that you can find acids in clouds, soil and even our own stomachs.

Energy issues are often covered in the news, so it was really useful to learn about different types of energy sources.

Hazam, *17 years old*

Get started

Get started with a short activity or discussion about the lesson.

Key terms

Key terms boxes give definitions of important words and phrases that you will come across.

Worked example

Worked examples provide a clear idea of what is required for a calculation.

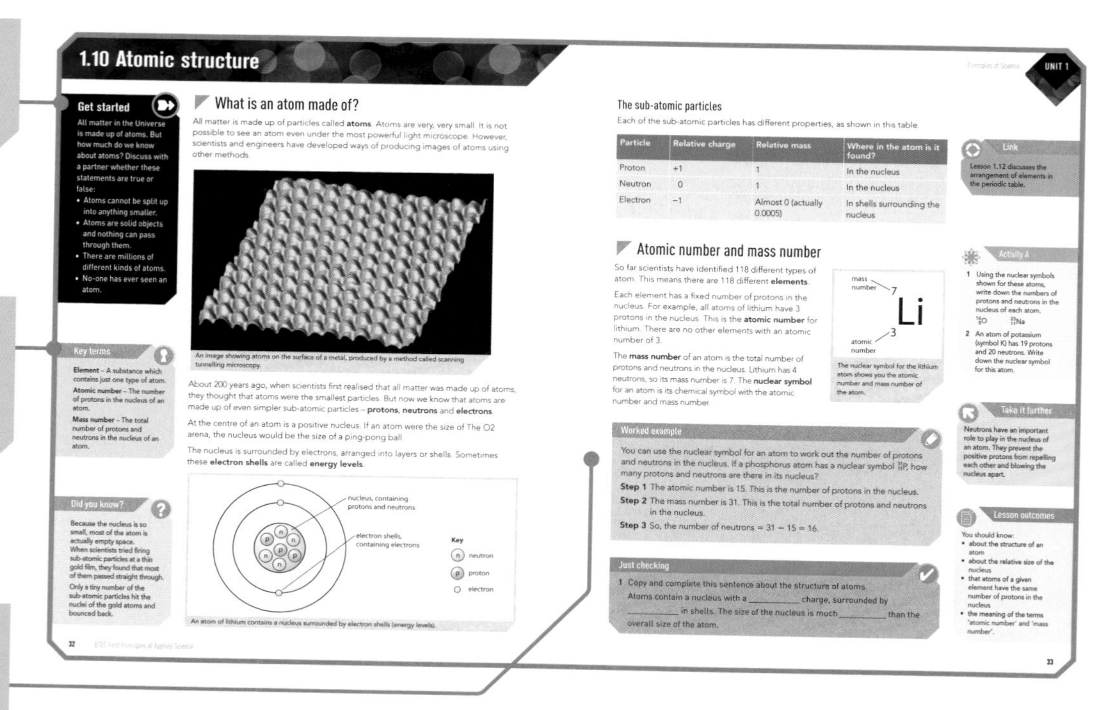

1.10 Atomic structure

Get started

All matter in the Universe is made up of atoms. But how much do we know about atoms? Discuss with a partner whether these statements are true or false:

- Atoms cannot be split up into anything smaller.
- Atoms are solid objects and nothing can pass through them.
- There are millions of different kinds of atoms.
- No-one has ever seen an atom.

What is an atom made of?

All matter is made up of particles called **atoms**. Atoms are very, very small. It is not possible to see an atom even under the most powerful light microscope. However, scientists and engineers have developed ways of producing images of atoms using other methods.

An image showing atoms on the surface of a metal, produced by a method called scanning tunnelling microscopy.

About 200 years ago, when scientists first realised that all matter was made up of atoms, they thought that atoms were the smallest particles. But now we know that atoms are made up of even simpler sub-atomic particles – **protons**, **neutrons** and **electrons**.

At the centre of an atom is a positive nucleus. If an atom were the size of The O2 arena, the nucleus would be the size of a ping-pong ball.

The nucleus is surrounded by electrons, arranged into layers or shells. Sometimes these **electron shells** are called **energy levels**.

An atom of lithium contains a nucleus surrounded by electron shells (energy levels).

Key terms

Element – A substance which contains just one type of atom.

Atomic number – The number of protons in the nucleus of an atom.

Mass number – The total number of protons and neutrons in the nucleus of an atom.

Did you know?

Because the nucleus is so small, most of the atom is actually empty space.
When scientists tried firing sub-atomic particles at a thin gold film, they found that most of them passed straight through.

Only a tiny number of the sub-atomic particles hit the nuclei of the gold atoms and bounced back.

The sub-atomic particles

Each of the sub-atomic particles has different properties, as shown in this table.

Particle	Relative charge	Relative mass	Where in the atom is it found?
Proton	+1	1	In the nucleus
Neutron	0	1	In the nucleus
Electron	−1	Almost 0 (actually 0.0005)	In shells surrounding the nucleus

Link

Lesson 1.12 discusses the arrangement of elements in the periodic table.

Atomic number and mass number

So far scientists have identified 118 different types of atom. This means there are 118 different **elements**.

Each element has a fixed number of protons in the nucleus. For example, all atoms of lithium have 3 protons in the nucleus. This is the **atomic number** for lithium. There are no other elements with an atomic number of 3.

The **mass number** of an atom is the total number of protons and neutrons in the nucleus. Lithium has 4 neutrons, so its mass number is 7. The **nuclear symbol** for an atom is its chemical symbol with the atomic number and mass number.

The nuclear symbol for the lithium atom shows you the atomic number and mass number of the atom.

Activity A

1 Using the nuclear symbols shown for these atoms, write down the numbers of protons and neutrons in the nucleus of each atom.
$^{16}_{8}O$ $^{23}_{11}Na$

2 An atom of potassium (symbol K) has 19 protons and 20 neutrons. Write down the nuclear symbol for this atom.

Worked example

You can use the nuclear symbol for an atom to work out the number of protons and neutrons in the nucleus. If a phosphorus atom has a nuclear symbol $^{31}_{15}P$, how many protons and neutrons are there in its nucleus?

Step 1 The atomic number is 15. This is the number of protons in the nucleus.

Step 2 The mass number is 31. This is the total number of protons and neutrons in the nucleus.

Step 3 So, the number of neutrons = 31 − 15 = 16.

Take it further

Neutrons have an important role to play in the nucleus of an atom. They prevent the positive protons from repelling each other and blowing the nucleus apart.

Just checking

1 Copy and complete this sentence about the structure of atoms.
Atoms contain a nucleus with a __________ charge, surrounded by __________ in shells. The size of the nucleus is much __________ than the overall size of the atom.

Lesson outcomes

You should know:

- about the structure of an atom
- about the relative size of the nucleus
- that atoms of a given element have the same number of protons in the nucleus
- the meaning of the terms 'atomic number' and 'mass number'.

You will be assessed in two different ways for your BTEC First Applied Science Award. For Unit 1 you will be assessed externally using a paper-based test. The Assessment Zone helps you to prepare for the test by showing you some of the different types of questions you will need to answer.

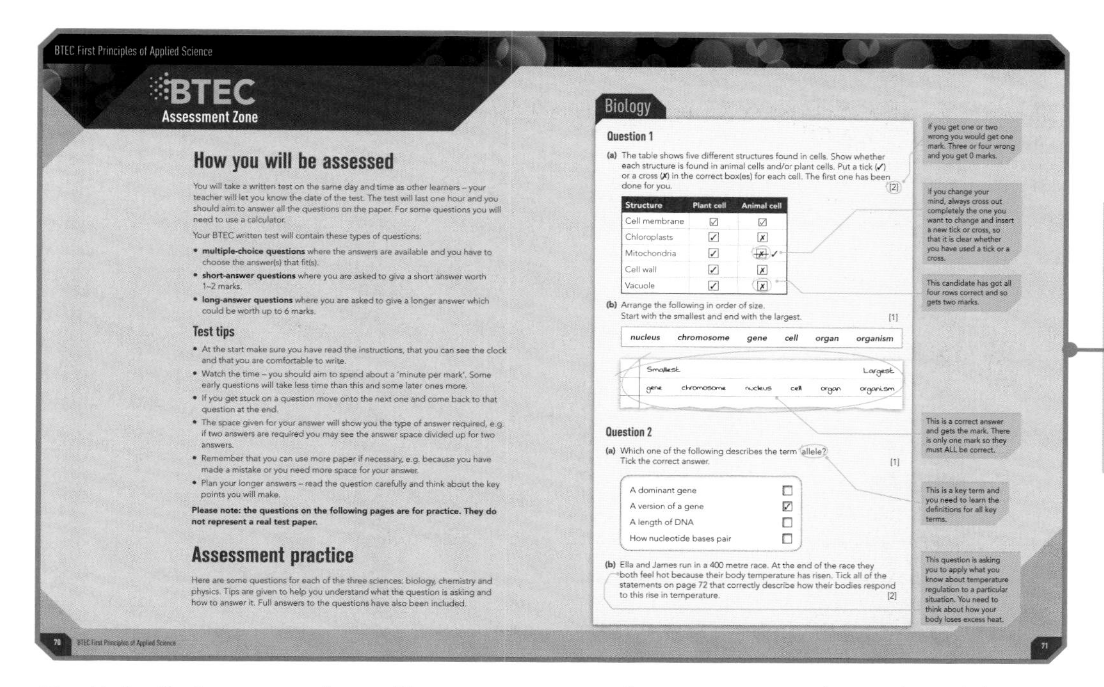

BTEC First Principles of Applied Science

BTEC Assessment Zone

How you will be assessed

You will take a written test on the same day and time as other learners – your teacher will let you know the date of the test. The test will last one hour and you should aim to answer all the questions on the paper. For some questions you will need to use a calculator.

Your BTEC written test will contain these types of questions:

- **multiple-choice questions** where the answers are available and you have to choose the answer(s) that fit(s).
- **short-answer questions** where you are asked to give a short answer worth 1–2 marks.
- **long-answer questions** where you are asked to give a longer answer which could be worth up to 6 marks.

Test tips

- At the start make sure you have read the instructions, that you can see the clock and that you are comfortable to write.
- Watch the time – you should aim to spend about a 'minute per mark'. Some early questions will take less time than this and some later ones more.
- If you get stuck on a question move onto the next one and come back to that question at the end.
- The space given for your answer will show you the type of answer required, e.g. if two answers are required you may see the answer space divided up for two answers.
- Remember that you can use more paper if necessary, e.g. because you have made a mistake or you need more space for your answer.
- Plan your longer answers – read the question carefully and think about the key points you will make.

Please note: the questions on the following pages are for practice. They do not represent a real test paper.

Assessment practice

Here are some questions for each of the three sciences: biology, chemistry and physics. Tips are given to help you understand what the question is asking and how to answer it. Full answers to the questions have also been included.

70 BTEC First Principles of Applied Science

Biology

Question 1

(a) The table shows five different structures found in cells. Show whether each structure is found in animal cells and/or plant cells. Put a tick (✓) or a cross (✗) in the correct box(es) for each cell. The first one has been done for you. [2]

Structure	Plant cell	Animal cell
Cell membrane	☑	☑
Chloroplasts	✓	✗
Mitochondria	✓	✗ ✓
Cell wall	✓	✗
Vacuole	✓	✗

(b) Arrange the following in order of size.
Start with the smallest and end with the largest. [1]

nucleus chromosome gene cell organ organism

Smallest … Largest
gene chromosome nucleus cell organ organism

Question 2

(a) Which one of the following describes the term allele?
Tick the correct answer. [1]

- A dominant gene ☐
- A version of a gene ☑
- A length of DNA ☐
- How nucleotide bases pair ☐

(b) Ella and James run in a 400 metre race. At the end of the race they both feel hot because their body temperature has risen. Tick all of the statements on page 72 that correctly describe how their bodies respond to this rise in temperature. [2]

If you get one or two wrong you would get one mark. Three or four wrong and you get 0 marks.

If you change your mind, always cross out completely the one you want to change and insert a new tick or cross, so that it is clear whether you have used a tick or a cross.

This candidate has got all four rows correct and so gets two marks.

This is a correct answer and gets the mark. There is only one mark so they must ALL be correct.

This is a key term and you need to learn the definitions for all key terms.

This question is asking you to apply what you know about temperature regulation to a particular situation. You need to think about how your body loses excess heat.

71

Assessment Zone

The Assessment Zone at the end of Unit 1 contains information on how you will be assessed. It also includes practice questions to help you prepare for your external test.

For Units 2–4 your teacher will set assignments for you to complete. The table in the BTEC Assessment Zone for Units 2–4 explains what you must do in order to achieve each of the assessment criteria.

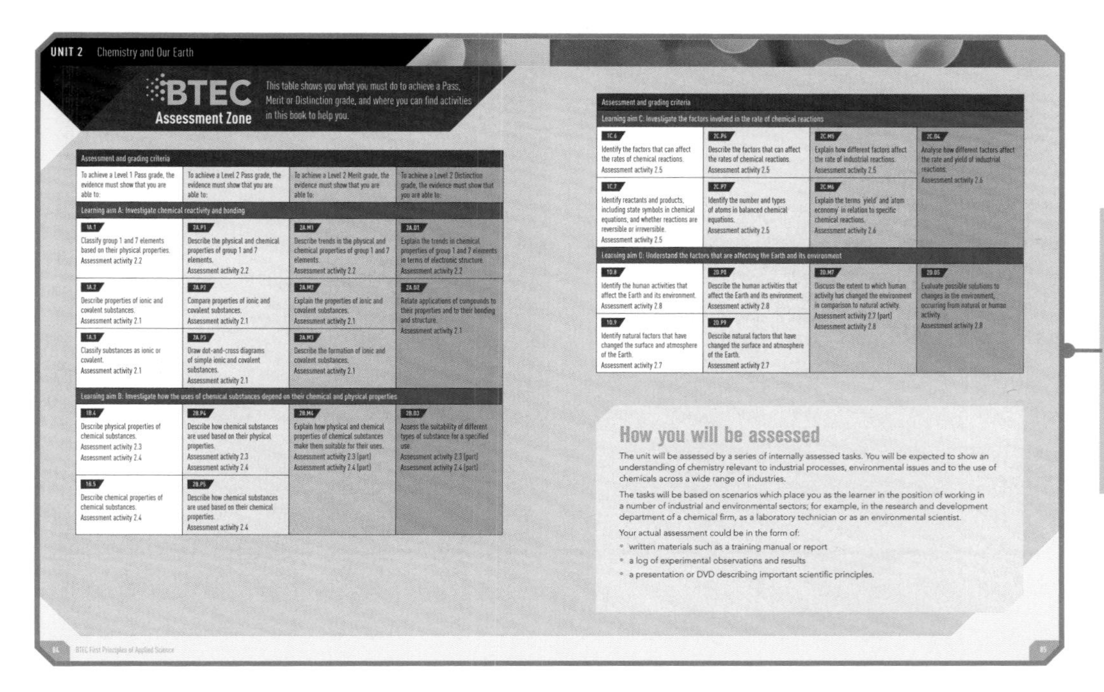

UNIT 2 Chemistry and Our Earth

BTEC Assessment Zone

This table shows you what you must do to achieve a Pass, Merit or Distinction grade, and where you can find activities in this book to help you.

Assessment and grading criteria			
To achieve a Level 1 Pass grade, the evidence must show that you are able to:	To achieve a Level 2 Pass grade, the evidence must show that you are able to:	To achieve a Level 2 Merit grade, the evidence must show that you are able to:	To achieve a Level 2 Distinction grade, the evidence must show that you are able to:
Learning aim A: Investigate chemical reactivity and bonding			
1A.1 Classify group 1 and 7 elements based on their physical properties. Assessment activity 2.2	2A.P1 Describe the physical and chemical properties of group 1 and 7 elements. Assessment activity 2.2	2A.M1 Describe trends in the physical and chemical properties of group 1 and 7 elements. Assessment activity 2.2	2A.D1 Explain the trends in chemical properties of group 1 and 7 elements in terms of electronic structure. Assessment activity 2.2
1A.2 Describe properties of ionic and covalent substances. Assessment activity 2.1	2A.P2 Compare properties of ionic and covalent substances. Assessment activity 2.1	2A.M2 Explain the properties of ionic and covalent substances. Assessment activity 2.1	2A.D2 Relate applications of compounds to their properties and to their bonding and structure. Assessment activity 2.1
1A.3 Classify substances as ionic or covalent. Assessment activity 2.1	2A.P3 Draw dot-and-cross diagrams of simple ionic and covalent substances. Assessment activity 2.1	2A.M3 Describe the formation of ionic and covalent substances. Assessment activity 2.1	
Learning aim B: Investigate how the uses of chemical substances depend on their chemical and physical properties			
1B.4 Describe physical properties of chemical substances. Assessment activity 2.3 Assessment activity 2.4	2B.P4 Describe how chemical substances are used based on their physical properties. Assessment activity 2.3 Assessment activity 2.4	2B.M4 Explain how physical and chemical properties of chemical substances make them suitable for their uses. Assessment activity 2.3 (part) Assessment activity 2.4 (part)	2B.D3 Assess the suitability of different types of substance for a specified use. Assessment activity 2.3 (part) Assessment activity 2.4 (part)
1B.5 Describe chemical properties of chemical substances. Assessment activity 2.4	2B.P5 Describe how chemical substances are used based on their chemical properties. Assessment activity 2.4		

Assessment and grading criteria			
Learning aim C: Investigate the factors involved in the rate of chemical reactions			
1C.6 Identify the factors that can affect the rates of chemical reactions. Assessment activity 2.5	2C.P6 Describe the factors that can affect the rates of chemical reactions. Assessment activity 2.5	2C.M5 Explain how different factors affect the rate of industrial reactions. Assessment activity 2.5	2C.D4 Analyse how different factors affect the rate and yield of industrial reactions. Assessment activity 2.6
1C.7 Identify reactants and products, including state symbols in chemical equations, and whether reactions are reversible or irreversible. Assessment activity 2.5	2C.P7 Identify the number and types of atoms in balanced chemical equations. Assessment activity 2.5	2C.M6 Explain the terms 'yield' and 'atom economy' in relation to specific chemical reactions. Assessment activity 2.6	
Learning aim D: Understand the factors that are affecting the Earth and its environment			
1D.8 Identify the human activities that affect the Earth and its environment. Assessment activity 2.8	2D.P8 Describe the human activities that affect the Earth and its environment. Assessment activity 2.8	2D.M7 Discuss the extent to which human activity has changed the environment in comparison to natural activity. Assessment activity 2.7 (part) Assessment activity 2.8	2D.D5 Evaluate possible solutions to changes in the environment, occurring from natural or human activity. Assessment activity 2.8
1D.9 Identify natural factors that have changed the surface and atmosphere of the Earth. Assessment activity 2.7	2D.P9 Describe natural factors that have changed the surface and atmosphere of the Earth. Assessment activity 2.7		

How you will be assessed

The unit will be assessed by a series of internally assessed tasks. You will be expected to show an understanding of chemistry relevant to industrial processes, environmental issues and to the use of chemicals across a wide range of industries.

The tasks will be based on scenarios which place you as the learner in the position of working in a number of industrial and environmental sectors; for example, in the research and development department of a chemical firm, as a laboratory technician or as an environmental scientist.

Your actual assessment could be in the form of:

- written materials such as a training manual or report
- a log of experimental observations and results
- a presentation or DVD describing important scientific principles.

Assessment and grading criteria

This table in the BTEC Assessment Zone signposts assessment activities you'll find in this book to help you to prepare for your assignments.

There are different types of activities for you to do to help you develop your knowledge, skills and understanding.

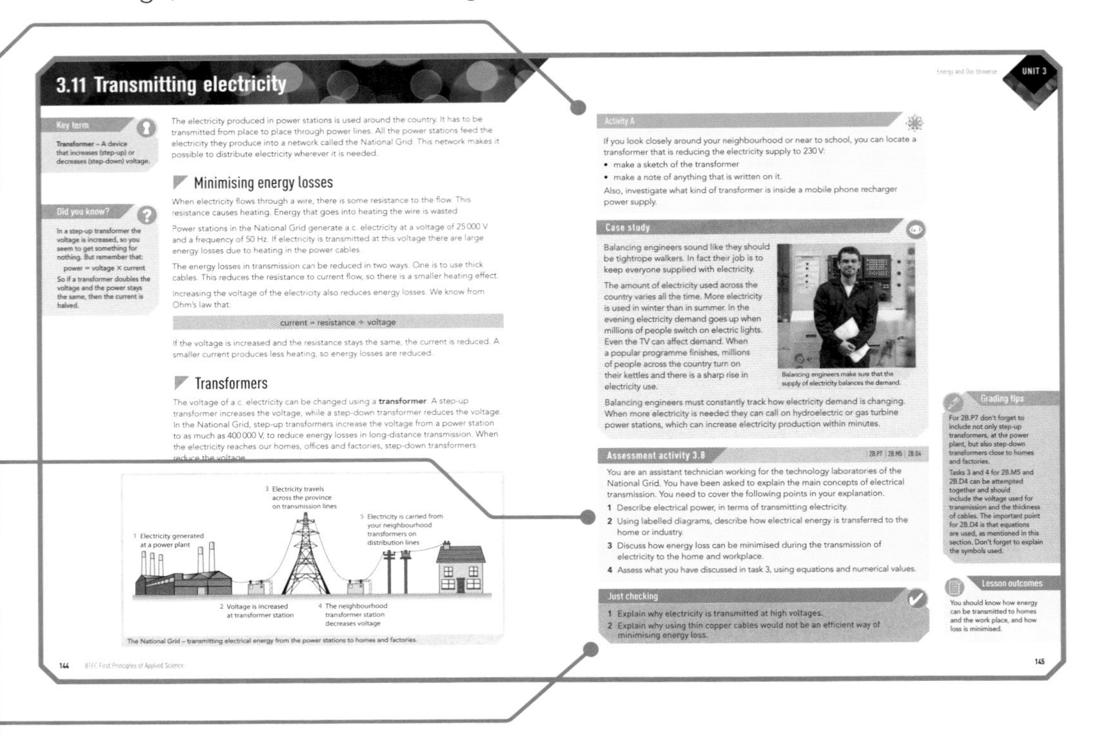
Energy and Our Universe UNIT 3

3.11 Transmitting electricity

Key term

Transformer – A device that increases (step-up) or decreases (step-down) voltage.

Did you know?

In a step-up transformer the voltage is increased, so you seem to get something for nothing. But remember that:

power = voltage × current

So if a transformer doubles the voltage and the power stays the same, then the current is halved.

The electricity produced in power stations is used around the country. It has to be transmitted from place to place through power lines. All the power stations feed the electricity they produce into a network called the National Grid. This network makes it possible to distribute electricity wherever it is needed.

Minimising energy losses

When electricity flows through a wire, there is some resistance to the flow. This resistance causes heating. Energy that goes into heating the wire is wasted.

Power stations in the National Grid generate a.c. electricity at a voltage of 25 000 V and a frequency of 50 Hz. If electricity is transmitted at this voltage there are large energy losses due to heating in the power cables.

The energy losses in transmission can be reduced in two ways. One is to use thick cables. This reduces the resistance to current flow, so there is a smaller heating effect.

Increasing the voltage of the electricity also reduces energy losses. We know from Ohm's law that:

current = resistance ÷ voltage

If the voltage is increased and the resistance stays the same, the current is reduced. A smaller current produces less heating, so energy losses are reduced.

Transformers

The voltage of a.c. electricity can be changed using a **transformer**. A step-up transformer increases the voltage, while a step-down transformer reduces the voltage. In the National Grid, step-up transformers increase the voltage from a power station to as much as 400 000 V, to reduce energy losses in long-distance transmission. When the electricity reaches our homes, offices and factories, step-down transformers reduce the voltage.

The National Grid – transmitting electrical energy from the power stations to homes and factories.

Activity A

If you look closely around your neighbourhood or near to school, you can locate a transformer that is reducing the electricity supply to 230 V:

- make a sketch of the transformer
- make a note of anything that is written on it.

Also, investigate what kind of transformer is inside a mobile phone recharger power supply.

Case study

Balancing engineers sound like they should be tightrope walkers. In fact their job is to keep everyone supplied with electricity.

The amount of electricity used across the country varies all the time. More electricity is used in winter than in summer. In the evening electricity demand goes up when millions of people switch on electric lights. Even the TV can affect demand. When a popular programme finishes, millions of people across the country turn on their kettles and there is a sharp rise in electricity use.

Balancing engineers make sure that the supply of electricity balances the demand.

Balancing engineers must constantly track how electricity demand is changing. When more electricity is needed they can call on hydroelectric or gas turbine power stations, which can increase electricity production within minutes.

Assessment activity 3.8 2B.P7 | 2B.M5 | 2B.D4

You are an assistant technician working for the technology laboratories of the National Grid. You have been asked to explain the main concepts of electrical transmission. You need to cover the following points in your explanation.

1 Describe electrical power, in terms of transmitting electricity.
2 Using labelled diagrams, describe how electrical energy is transferred to the home or industry.
3 Discuss how energy loss can be minimised during the transmission of electricity to the home and workplace.
4 Assess what you have discussed in task 3, using equations and numerical values.

Grading tips

For 2B.P7 don't forget to include not only step-up transformers, at the power plant, but also step-down transformers close to homes and factories.

Tasks 3 and 4 for 2B.M5 and 2B.D4 can be attempted together and should include the voltage used for transmission and the thickness of cables. The important point for 2B.D4 is that equations are used, as mentioned in this section. Don't forget to explain the symbols used.

Just checking

1 Explain why electricity is transmitted at high voltages.
2 Explain why using thin copper cables would not be an efficient way of minimising energy loss.

Lesson outcomes

You should know how energy can be transmitted to homes and the work place, and how loss is minimised.

144 BTEC First Principles of Applied Science 145

Activities

Activities will help you show the knowledge and understanding you have gained in the lesson.

Assessment activities

Assessment activities are suggestions for tasks that you might do to help build towards your assignment. Each Assessment activity has **Grading tips** which provide guidance on how to achieve the best grade.

Just checking

Use these to check your knowledge and understanding at the end of the lesson.

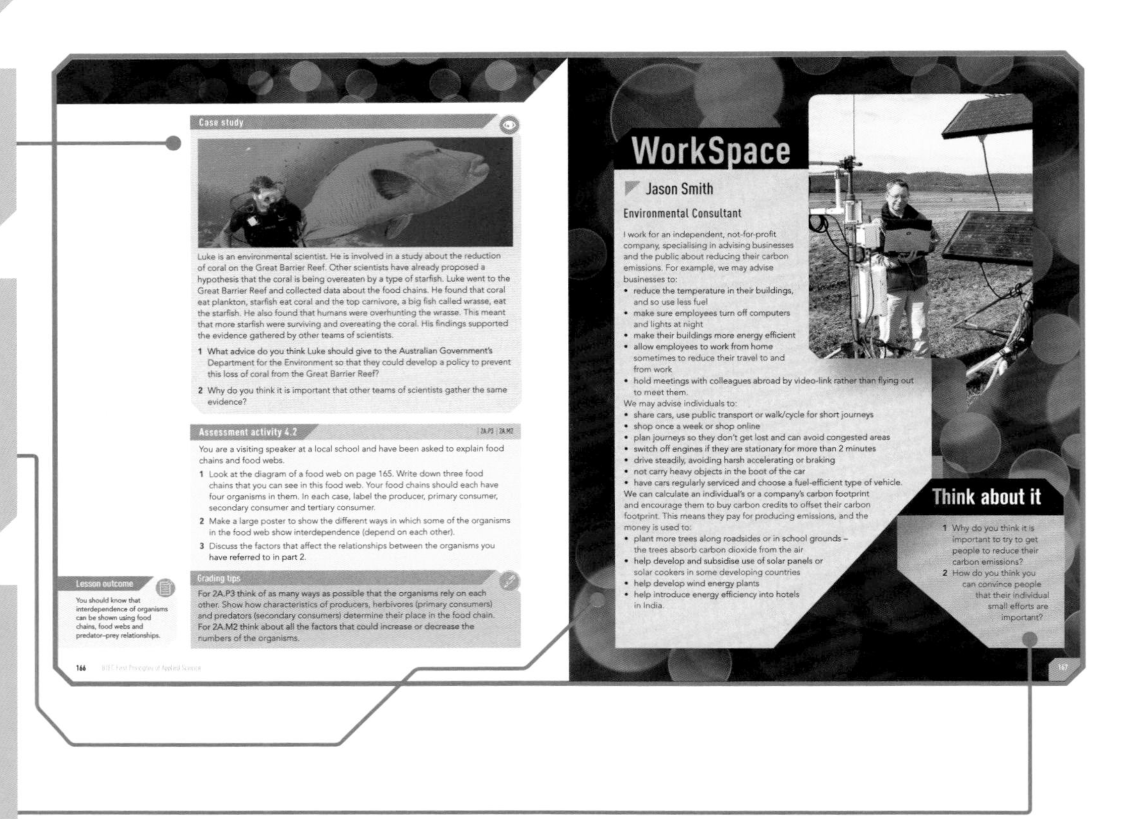
Case study

Luke is an environmental scientist. He is involved in a study about the reduction of coral on the Great Barrier Reef. Other scientists have already proposed a hypothesis that the coral is being overeaten by a type of starfish. Luke went to the Great Barrier Reef and collected data about the food chains. He found that coral eat plankton, starfish eat coral and the top carnivore, a big fish called wrasse, eat the starfish. He also found that humans were overhunting the wrasse. This meant that more starfish were surviving and overeating the coral. His findings supported the evidence gathered by other teams of scientists.

1 What advice do you think Luke should give to the Australian Government's Department for the Environment so that they could develop a policy to prevent this loss of coral from the Great Barrier Reef?
2 Why do you think it is important that other teams of scientists gather the same evidence?

Assessment activity 4.2 2A.P3 | 2A.M2

You are a visiting speaker at a local school and have been asked to explain food chains and food webs.

1 Look at the diagram of a food web on page 165. Write down three food chains that you can see in this food web. Your food chains should each have four organisms in them. In each case, label the producer, primary consumer, secondary consumer and tertiary consumer.
2 Make a large poster to show the different ways in which some of the organisms in the food web show interdependence (depend on each other).
3 Discuss the factors that affect the relationships between the organisms you have referred to in part 2.

Lesson outcome

You should know that interdependence of organisms can be shown using food chains, food webs and predator-prey relationships.

Grading tips

For 2A.P3 think of as many ways as possible that the organisms rely on each other. Show how characteristics of producers, herbivores (primary consumers) and predators (secondary consumers) determine their place in the food chain. For 2A.M2 think about all the factors that could increase or decrease the numbers of the organisms.

WorkSpace

Jason Smith

Environmental Consultant

I work for an independent, not-for-profit company, specialising in advising businesses and the public about reducing their carbon emissions. For example, we may advise businesses to:

- reduce the temperature in their buildings, and so use less fuel
- make sure employees turn off computers and lights at night
- make their buildings more energy efficient
- allow employees to work from home sometimes to reduce their travel to and from work
- hold meetings with colleagues abroad by video-link rather than flying out to meet them.

We may advise individuals to:

- share cars, use public transport or walk/cycle for short journeys
- shop once a week or shop online
- plan journeys so they don't get lost and can avoid congested areas
- switch off engines if they are stationary for more than 2 minutes
- drive steadily, avoiding harsh accelerating or braking
- not carry heavy objects in the boot of the car
- have cars regularly serviced and choose a fuel-efficient type of vehicle.

We can calculate an individual's or a company's carbon footprint and encourage them to buy carbon credits to offset their carbon footprint. This means they pay for producing emissions, and the money is used to:

- plant more trees along roadsides or in school grounds – the trees absorb carbon dioxide from the air
- help develop and subsidise use of solar panels or solar cookers in some developing countries
- help develop wind energy plants
- help introduce energy efficiency into hotels in India.

Think about it

1 Why do you think it is important to try to get people to reduce their carbon emissions?
2 How do you think you can convince people that their individual small efforts are important?

166 BTEC First Principles of Applied Science 167

Case studies

Case studies show you examples of specific scientific topics in the workplace.

WorkSpace

WorkSpaces provide a snapshot of someone who works in the industry and of real workplace issues. They show how the skills and knowledge you develop can help you in your career.

Think about it

Workspaces also give you the chance to think more about the role that this person does, and whether you want follow in their footsteps once you've completed your BTEC.

Planning and getting organised

The first step in managing your time is to plan ahead and be well organised. You can improve your planning and organisational skills by:

- Using a diary to schedule all the work you have to do. You could use this as a 'to do' list and tick off each task as you go.
- Dividing up long or complex tasks into manageable chunks and putting each 'chunk' in your diary with a deadline of its own.
- Always allowing more time than you think you need for a task.

Organising and selecting information

Once you have gathered a range of information during your research, you will need to organise the information so it's easy to use.

- Make sure your written notes are neat and have a clear heading – it's often useful to date them, too.
- Always keep a note of where the information came from (the title of a book, the title and date of a newspaper or magazine and the web address of a website) and, if relevant, which pages.
- Work out the results of any questionnaires you've used.

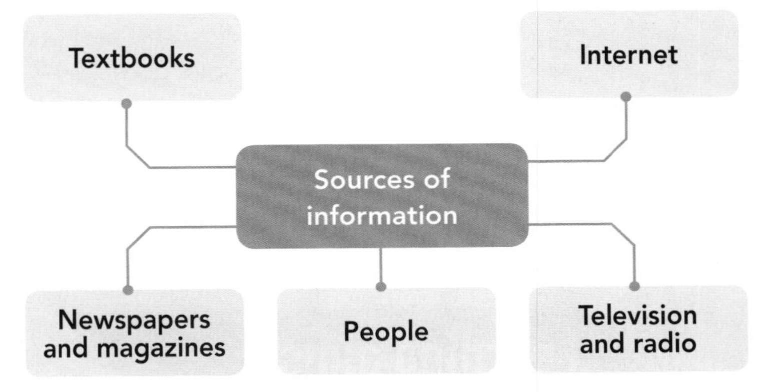

Once you have completed your research, re-read the assignment brief or instructions you were given to remind yourself of the exact wording of the question(s) and divide your information into three groups:

- Information that is totally relevant.
- Information that is not as good, but which could come in useful.
- Information that doesn't match the questions or assignment brief very much, but that you kept because you couldn't find anything better!

Check that there are no obvious gaps in your information against the questions or assignment brief. If there are, make a note of them so that you know exactly what you still have to find.

Presenting your work

Before handing in any assignments, make sure:

- you have addressed each part of the question and that your work is as complete as possible
- all spelling and grammar is correct
- you have referenced all sources of information you used for your research
- that all work is your own – otherwise you could be committing **plagiarism**
- you have saved a copy of your work.

Introduction

What use are applied science lessons?

Well – quite a lot of use actually. Look around the modern world and you will see science in action everywhere you turn.

The first part of this unit looks at the many different types of cell in your body. The knowledge gained from cell research has led to medically useful substances such as vaccines and antibodies. This unit also looks at DNA. DNA fingerprinting is used in every police CSI lab, and methods for modifying DNA are at the centre of public debates about cloning and genetically modified crops.

The second part of this unit explores the structure of atoms and what makes each type of atom different. Without this knowledge we would not know how to make light metal alloys for aircraft, or semiconductors for electronics. This section also discusses neutralisation reactions. These chemical reactions are used, for example, to make poor soils into productive farmland.

The final part of this unit is about forms of energy and how they can be changed, stored and used. This is the science behind producing electricity, heating our homes, and powering cars and other vehicles. You will also learn about electromagnetic waves. TVs, mobile phones, Facebook, Twitter – in fact, anything involving communications – rely on electromagnetic waves.

Assessment: You will be assessed externally using a paper-based test.

Learning aims

After completing this unit you should know more about:

a cells, organs and genes
b the roles of the nervous and endocrine systems in homeostasis and communication
c atomic structure and the periodic table
d substances and chemical reactions
e the importance of energy stores, energy transfers and energy transformations
f the properties and applications of waves in the electromagnetic spectrum.

The biology unit was really interesting. We got to look at our own cheek cells under a microscope, as well as plant and animal cells.

In the chemistry section we found out about acids and how they can be neutralised – and that you can find acids in clouds, soil and even our own stomachs.

Energy issues are often covered in the news, so it was really useful to learn about different types of energy sources.

Hazam, *17 years old*

Principles of Science

1

1.1 Cells – structure and function

Key term 

Organelles – Small structures within cells. Each carries out a particular function.

How are we made?

All living things on Earth are made of **cells**. Some consist of just one cell but some, like you, are made of many cells. Cells are the building blocks of organisms. Cells are very small but you can see them with a microscope.

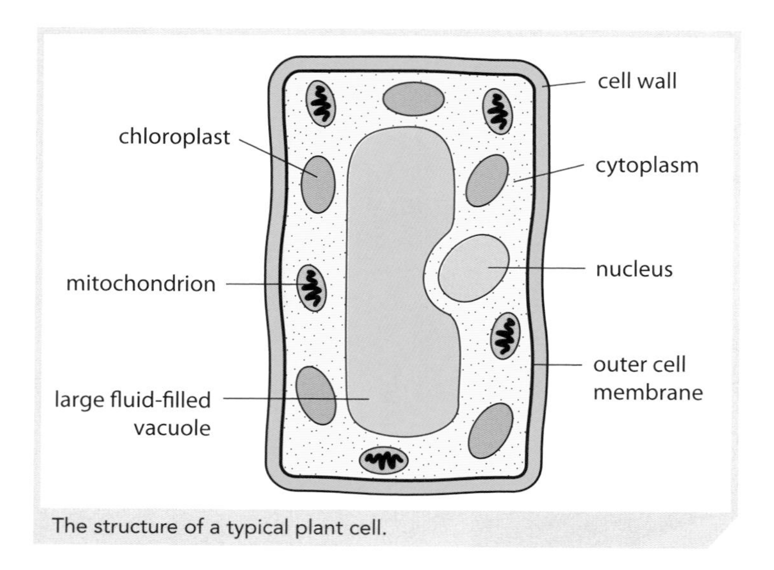

The structure of a typical plant cell.

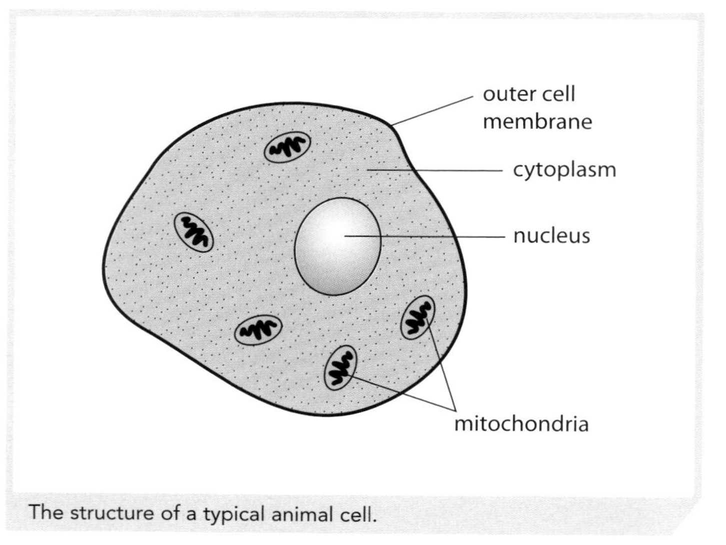

The structure of a typical animal cell.

Did you know? 

Plant and animal cells are called eukaryotic cells. Eukaryotic organisms have cells with their genetic material enclosed in a nucleus. Nearly all of the living things on Earth, except bacteria, are eukaryotes.

Case study 

Sue is a cytology screener. She works in a hospital laboratory and has to look at microscope slides of cells to see if any are abnormal. This could indicate that the person whose cells they are has early stage **cancer**. Sue has to know about cell structure so she can tell abnormal cells from normal ones. Cytology is the study of cell structure.

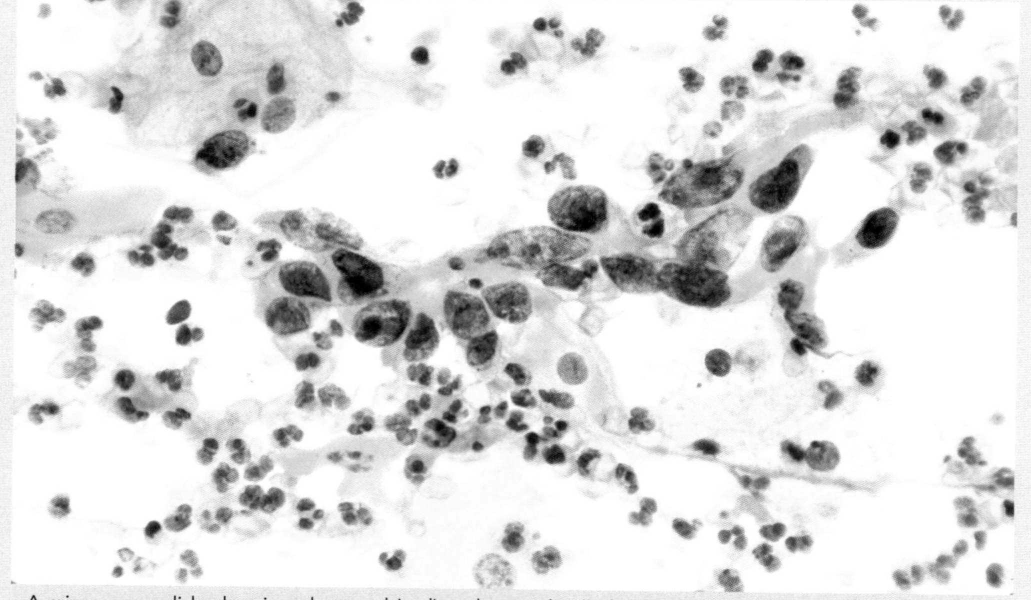

A microscope slide showing abnormal (red) and normal cells from the cervix.

Link 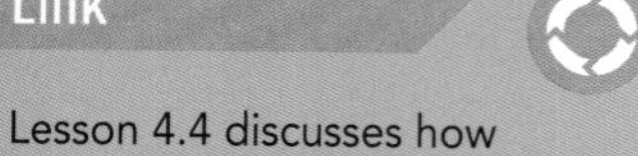

Lesson 4.4 discusses how viruses are *not* made of cells.

Plant and animal cells have many components, called **organelles**, such as **chloroplasts** and mitochondria. The table shows the different functions of the organelles found in plant and animal cells.

Cell component	Function	In plant cells?	In animal cells?
Outer cell membrane	Encloses the cell and keeps it separate from its environment. It controls how substances enter and leave the cell.	✓	✓
Nucleus	Contains DNA (genetic information) arranged into chromosomes. It controls the activities of the cell.	✓	✓
Cytoplasm	Jelly-like substance in which the organelles are held. Many chemical reactions take place in the cytoplasm.	✓	✓
Mitochondrion (plural: mitochondria)	The organelle where **aerobic respiration** – the release of energy from glucose or fat in the presence of oxygen – takes place.	✓	✓
Cell wall	Made from **cellulose**. The cell wall is very strong and prevents the cell from bursting. It gives the cell shape.	✓	✗
Chloroplast	Contains **chlorophyll**, which absorbs light energy for photosynthesis.	✓	✗
Vacuole	Contains cell sap and provides extra support for the cell. The vacuole can store nutrients and waste products.	✓	✗

Activity A

List the ways that animal and plant cells are **(a)** similar to each other and **(b)** different from each other.

Common misconception

Many people think that plants do not have genes. Their cells contain a nucleus, with DNA organised into chromosomes. Genes are lengths of DNA. So plants do have genes.

Just checking

For each of the following statements, say whether it is true or false.

1 All living things are made of many eukaryotic cells.
2 Plant cells do not contain mitochondria.
3 Animal cells do not have a large vacuole.
4 The cell membrane controls which substances enter and leave a cell.

Lesson outcomes

You should know about the basic structure and function of some eukaryotic cells and of some of their components.

1.2 Specialised cells

You, like many other organisms including plants, started life as a single cell – a fertilised egg. This divides and forms an **embryo**. Cells become **specialised** to perform different functions. This is called **differentiation** (becoming different).

Some examples of specialised cells are shown below. **(a)** to **(f)** are animal cells and **(g)** to **(j)** are plant cells. They all have the same basic components, but each type has some differences in structure. These differences, also called **adaptations**, enable the cells to carry out their particular function (job). Each type of cell is adapted for its function.

(a) Egg cell – for sexual reproduction.

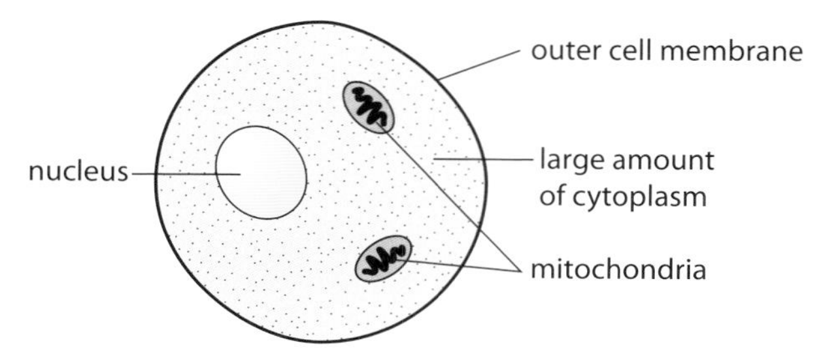

(b) Sperm cell – for sexual reproduction. The sperm has a tail to help it move to find the egg. It also has a large number of mitochondria to supply the energy needed for the movement.

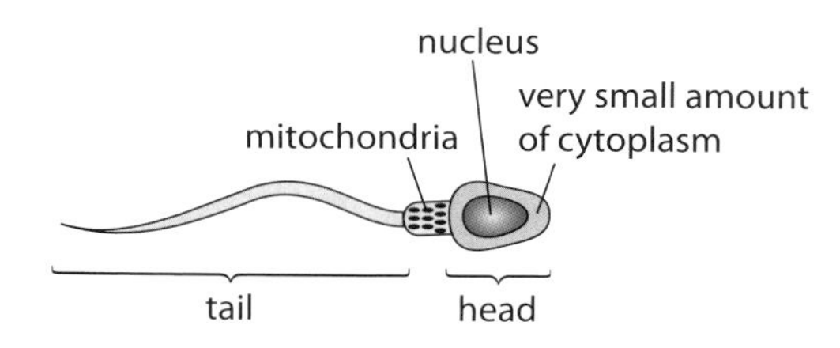

(c) **Red blood cells** – small cells, packed with haemoglobin which carries oxygen. They have no nucleus, no mitochondria and little cytoplasm, in order to carry more oxygen.

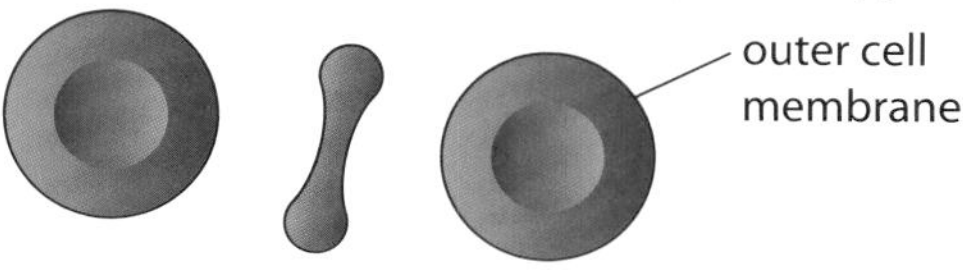

(d) **White blood cells** – for defence against infection. There are different types of white blood cell, but they all have a nucleus, cell membrane, cytoplasm and mitochondria.

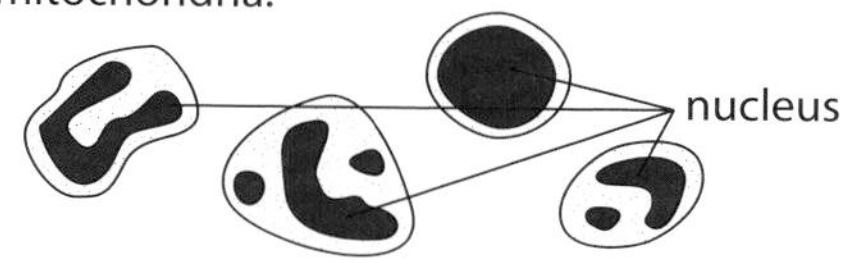

(e) Sensory neurone – long fibres carry impulses from sensory receptors to brain and spinal cord.

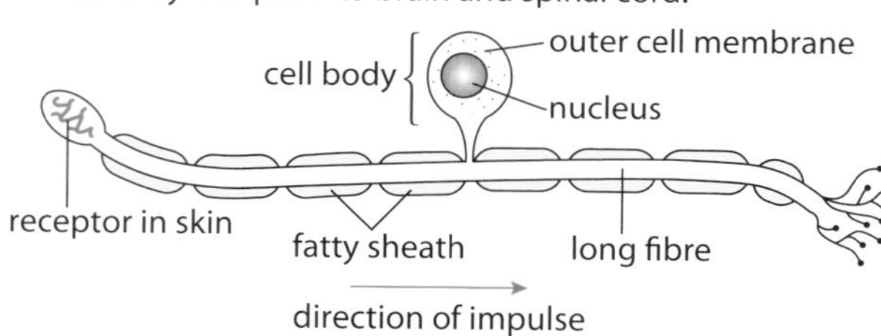

(f) Motor neurone – long fibres carry impulses from brain or spinal cord to muscles or glands.

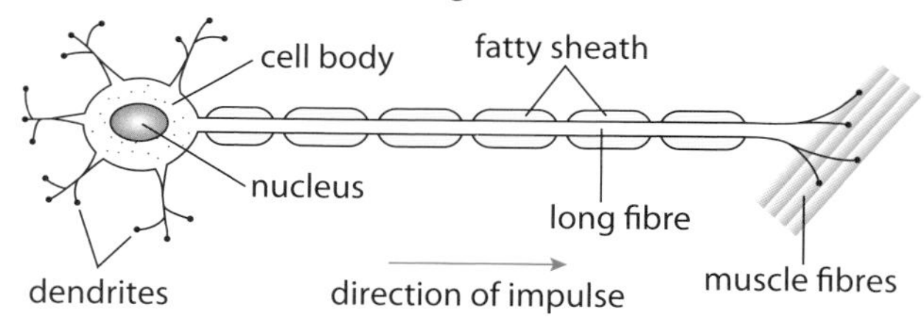

(g) Root hair cell – for absorption.

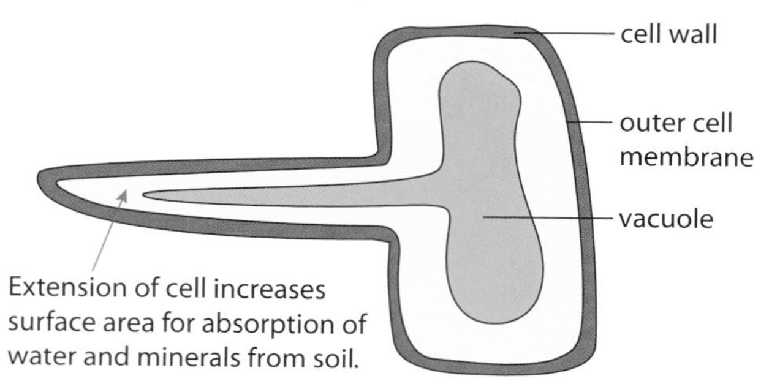

(h) Xylem – to carry water. There is no cytoplasm or organelles.

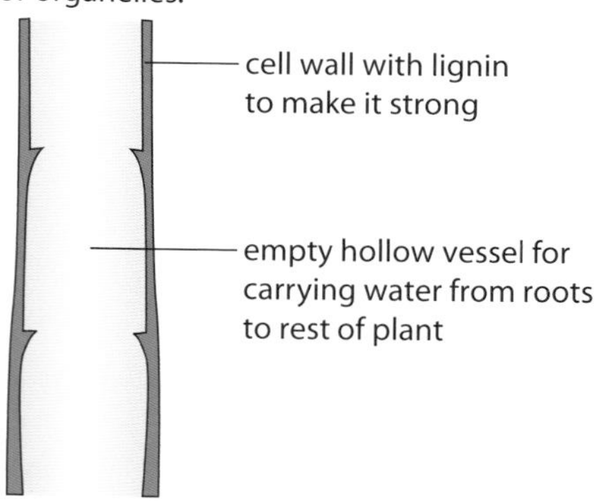

i) Phloem – carries dissolved sugars from leaves to other parts of plant.

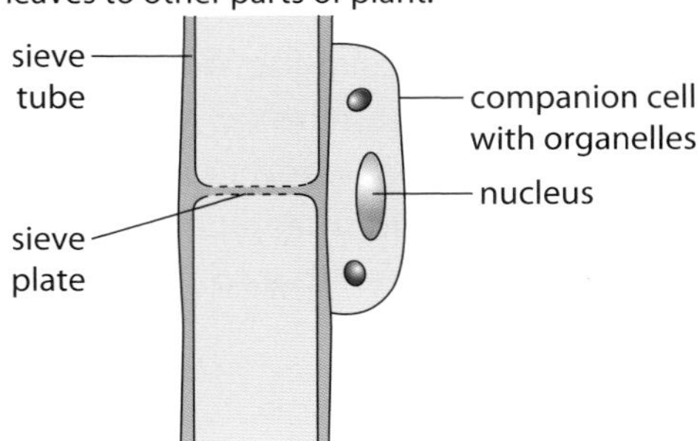

(j) Guard cells in epidermis of leaf – to allow carbon dioxide to enter for photosynthesis.

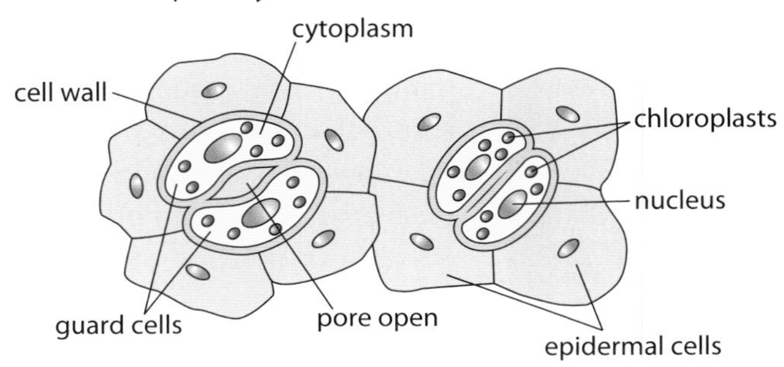

Tissues

A tissue is a group of similar, specialised cells that work together to carry out a particular function in the body. Some examples of animal tissues are given in the table below.

Type of tissue	Where found in the body	Functions
Epithelial (covering tissues)	Outside of skin. Make up the inside linings of heart, blood vessels, airways and gut. Some epithelial cells have tiny hairs, called cilia, on them.	Protection, absorption, filtration and **excretion**.
	Glands are also made of epithelial cells.	Production of useful substances, such as saliva, **hormones** and mucus.
Connective	Bone	Makes up skeleton – supports you and allows you to move.
	Blood	Transport and defence against invading **microorganisms**.
Muscle	Smooth muscle in walls of blood vessels and gut	Able to contract – arteries pulsate. Gut moves food through it.
	Heart muscle	Heart beats to pump blood around body.
Nervous	Nerves, spinal cord and brain	For sensing and responding in a coordinated way to the environment.

Activity A

1 Draw large, clear diagrams of one specialised animal cell and one specialised plant cell.

2 Label and annotate (write notes next to the labels) the diagrams to show how each cell is adapted for its function. You should also research, using the Internet and textbooks, to find out more about how the cells you have chosen are adapted. Use a ruler to draw the label lines.

Just checking

For each of the following statements, say whether it is true or false.

1 Red blood cells cannot carry out aerobic respiration because they do not have mitochondria.

2 Neurones are the longest cells in your body.

3 Blood and bone are types of connective tissue.

Lesson outcomes

You should know how cells become adapted to carry out specific functions, and that cells form tissues.

1.3 Organs and organ systems

Get started

Think of all the organs you have in your body. You have got different body systems, like the digestive system which is for digesting food. What are the following systems for:

- cardiovascular system
- respiratory system
- immune system
- nervous system?

Did you know?

Your brain accounts for 3% of your body mass. However, its cells are always busy and your brain uses 17% of your total energy intake every day.

Take it further

Research and find out about *one* of the following body systems. Describe the bodily function it carries out and the organs that are in it:

- immune system
- respiratory system
- nervous system
- endocrine system
- digestive system
- reproductive system
- integumentary (skin system)
- excretory system
- musculoskeletal system.

Activity A

The cardiovascular system may also be called the circulatory system. Why is this a good alternative name?

Organs

An organ is a group of tissues working together to do a particular job. You have many organs in your body and some do many jobs. The diagram below shows some of your organs and their functions. Doctors and surgeons need to know the positions of organs and what they do in order to diagnose and treat illness. Nurses, physiotherapists and other people working in medicine and health also need a working knowledge of the organ systems.

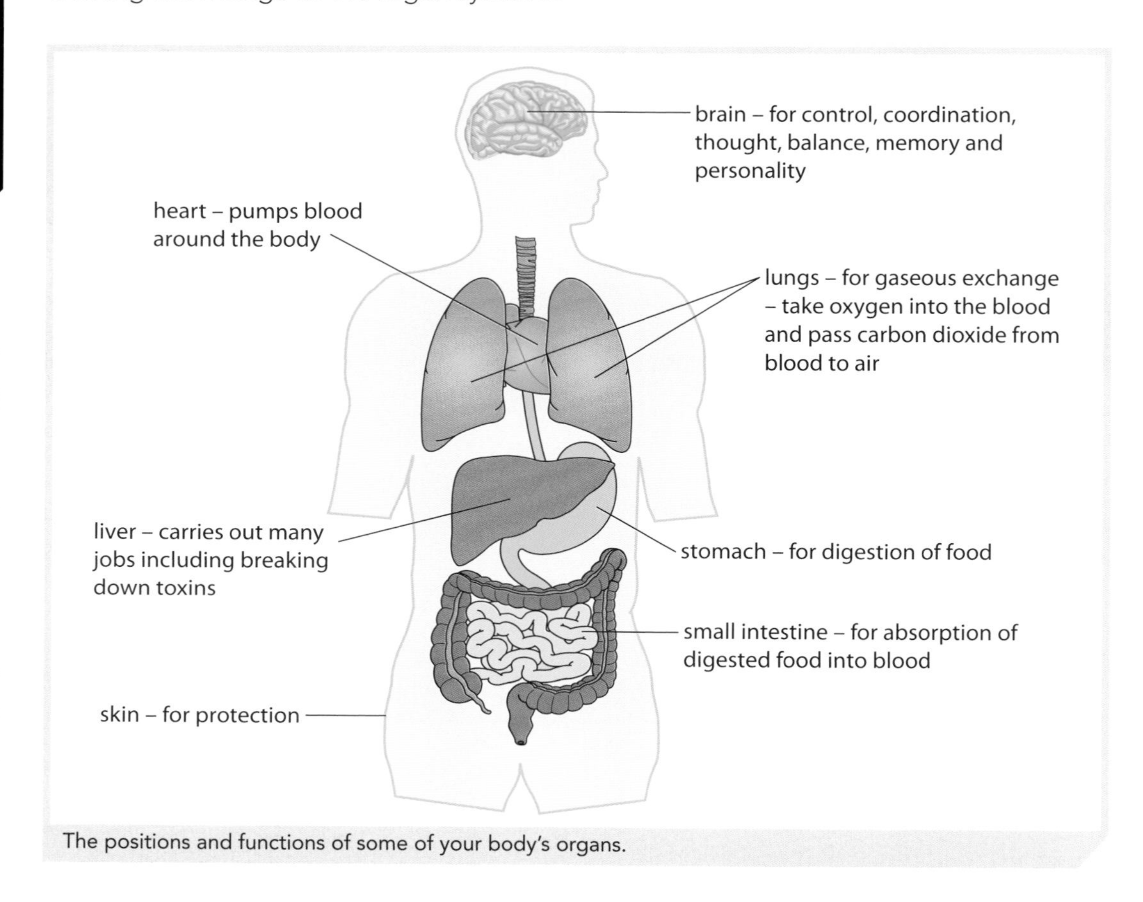

The positions and functions of some of your body's organs.

Organ systems

Organ systems are groups of organs that work together to carry out a particular task. For example, the **cardiovascular system** consists of the heart, **arteries**, **veins** and blood. It is the body's main transport network. It carries oxygen and nutrients to the body cells, and transports away carbon dioxide and other wastes.

Your heart is a muscular pump. It beats about 70 times a minute every day.

- Blood is pumped away from the heart in arteries and travels back to the heart in veins.
- When arteries reach the body tissues, they divide into smaller vessels called capillaries. The capillaries then join up to form veins.
- Substances pass into and out of the blood at the capillaries.

Plants have organs too

Plant cells are also organised into organs and tissues, such as leaves, roots, xylem and phloem.

Discussion point

In small groups discuss why your existence depends on leaves.

Leaves

Your whole existence depends on leaves. These are the plant organs where **photosynthesis** takes place. The cells in the mesophyll tissue in leaves have chloroplasts containing chlorophyll. They can trap the energy from sunlight and use it, plus water (from soil) and carbon dioxide (from air) to make the carbohydrate called glucose.

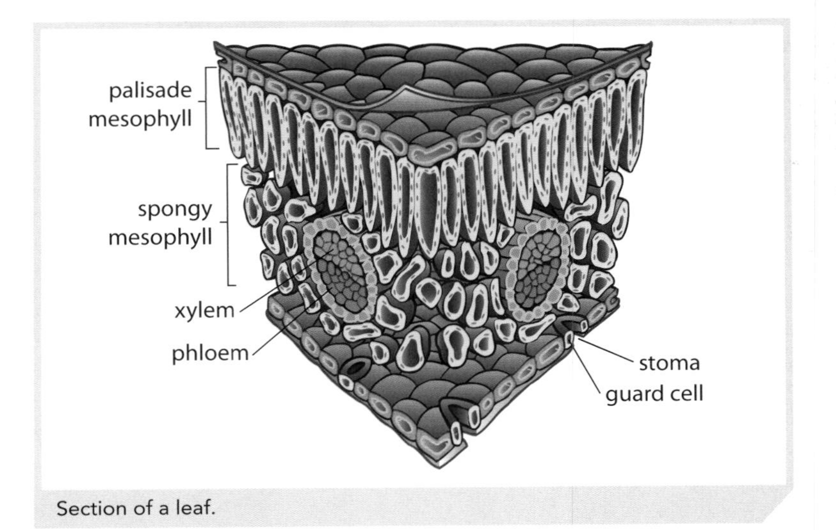

Section of a leaf.

Roots

These organs anchor the plant in soil. They also take up water and mineral salts from the soil.

Xylem and phloem

Xylem carries water and mineral salts from the roots to the leaves. Leaves need water for photosynthesis. Phloem carries the dissolved food made by leaves (mainly sugar) to other parts of the plant.

Transpiration

Look at the picture of guard cells in lesson 1.2. When the pores are open the leaves can take in carbon dioxide for photosynthesis. Another function of these pores (their proper name is **stomata**) is that when they are open, water vapour is lost from the leaves. This is known as transpiration. This pulls water up, from the roots to the leaves, in the xylem vessels.

Just checking

1 For each of the following, state which organ system it belongs to: stomach, brain, optic nerve, lung, heart, small intestine.

2 Match the following plant organs with a correct function.

Organ	Function
leaf	anchors plant in the soil
stem	carries out photosynthesis
root	holds plant upwards so leaves can trap energy from sunlight

3 Which one of the following describes transpiration?
- movement of water up xylem vessels
- opening pores (stomata) on leaves
- loss of water vapour from leaves

Lesson outcomes

You should know:
- that tissues form organs and organs work together to form systems
- the functions of leaves, roots, xylem and phloem
- the mechanism of transpiration.

1.4 DNA and chromosomes

Key term

Gene – A section of DNA that carries the instructions for a characteristic.

Genetic material

Anyone working in **forensic science** needs to understand the structure of DNA. By the 1940s, scientists knew that DNA (deoxyribonucleic acid), found in the **chromosomes**, was the genetic material. However, they did not know *how* it carried the instructions for different characteristics. They had to find out its molecular structure.

By 1953, James Watson and Francis Crick, with valuable help from Rosalind Franklin, had worked out that structure.

The double helix

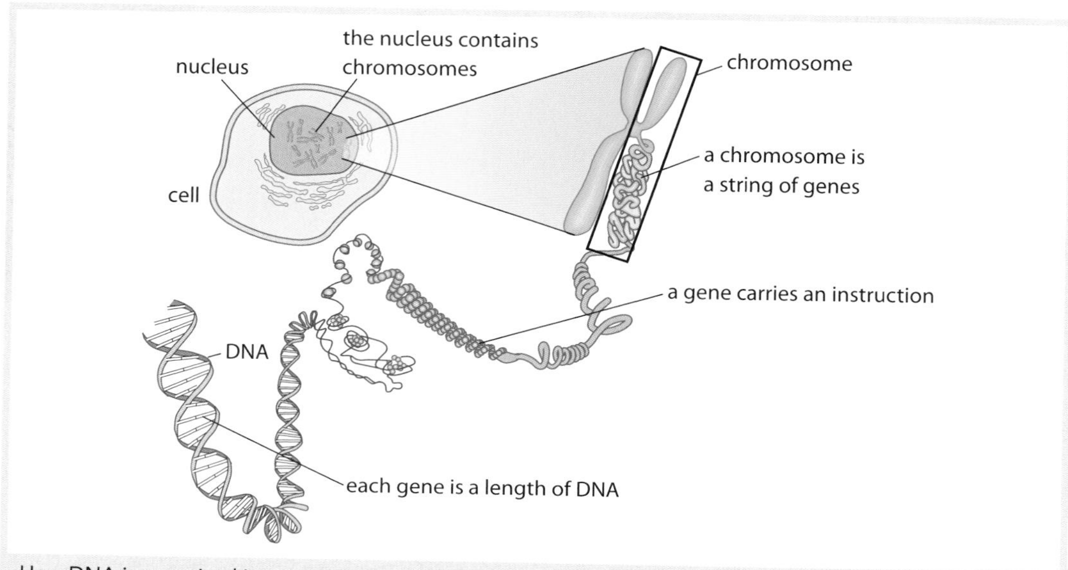

How DNA is organised into genes and chromosomes in the nucleus of a cell.

DNA is a very large molecule made up of two strands that twist together to form a **double helix**. The two strands are linked together by chemicals called bases. There are four different bases – **adenine (A)**, **thymine (T)**, **guanine (G)** and **cytosine (C)**.

The bases are joined and act like rungs on a ladder. You can see from the diagram below that A always pairs with T and G always pairs with C. This pairing is called **complementary base pairing**.

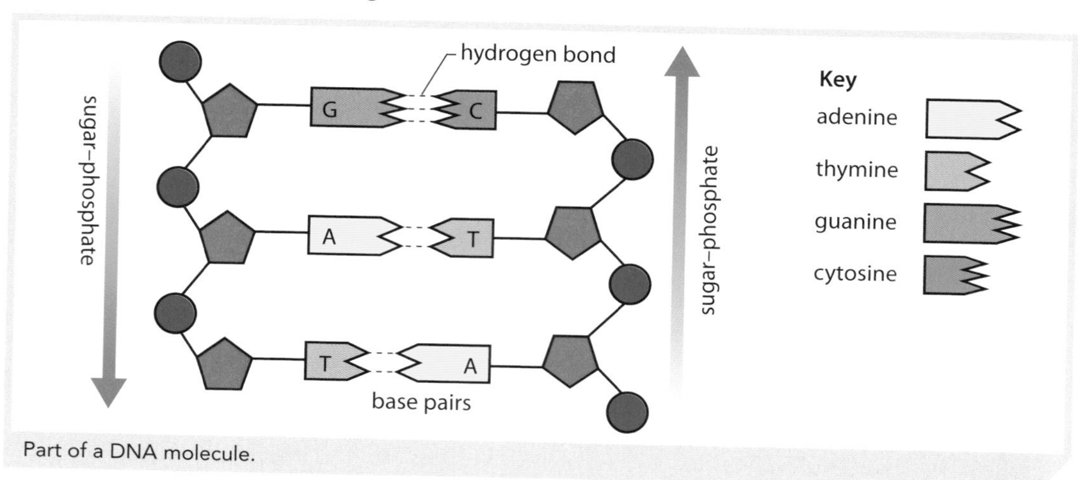

Part of a DNA molecule.

Activity A

A scientist analysed a short piece of DNA and found that it contained 100 base pairs (pairs of bases), and 18 of the bases were adenine. How many were **(a)** thymine, **(b)** guanine and **(c)** cytosine?

Chromosomes

In human cells, the DNA is organised into chromosomes, which are in the nucleus. We have 23 pairs of chromosomes in our body cells. You inherited one of each pair from your mother and one from your father.

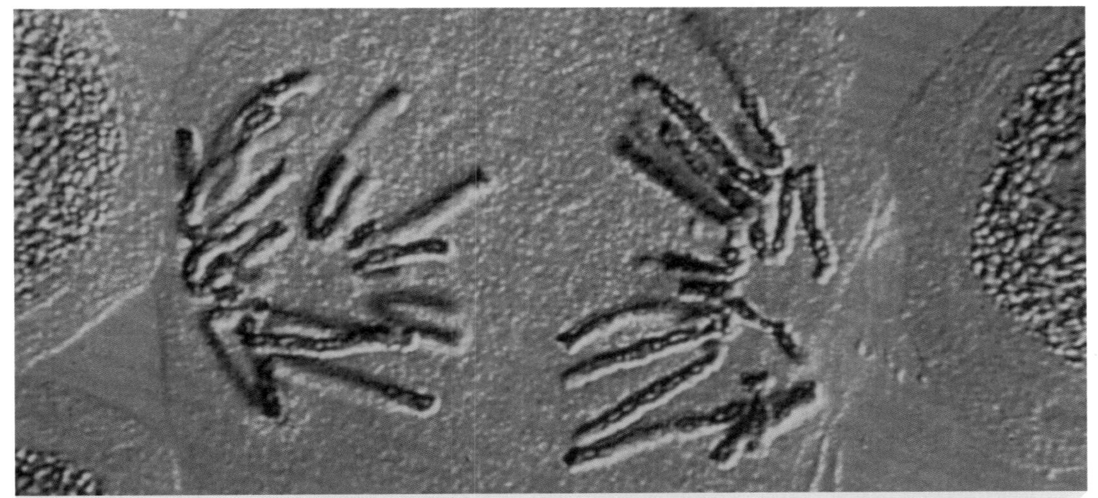

Bluebell cell dividing. You can see the chromosomes. Notice that the chromosomes are in pairs. This is so that when the cell divides into two, each new cell gets an identical copy of every chromosome.

Did you know?

A **mutation** is a change to genetic material, either to genes or to chromosomes. Some mutations change the sequence of bases in a length of DNA (a gene). This change means that the gene now has a slightly different code and will change the characteristic it codes for.

Activity B

How many chromosomes are there in these human cells:

1 a skin cell
2 a brain cell
3 a kidney cell
4 an egg cell
5 a red blood cell
6 a liver cell
7 a bone cell
8 an embryo cell?

Did you know?

We each have about 20 000 genes packaged into the chromosomes. All your genes make up your genome. Altogether, you have about 3 billion base pairs in your genome. Their sequence has been worked out by scientists working on The Human Genome Project. If you were to read the sequence aloud, non-stop, it would take 9.5 years.

Genes

Shorter lengths of DNA, within a chromosome, are called **genes**. Each gene contains a *specific sequence of bases*, which is the coded instructions for a characteristic. Different versions of the same gene are called **alleles**.

Homozygous and heterozygous

For every gene you inherit two alleles – one from each parent. The members of each pair of chromosomes contain the same genes but may have different alleles.

- If both alleles are the same for a particular gene, you are **homozygous** (*homo-* means 'the same') for that characteristic.
- If the two alleles for a particular gene are different you are **heterozygous** (*hetero-* means 'different') for that characteristic.

These two words, homozygous and heterozygous, describe your **genotype** for a particular characteristic.

Lesson outcomes

You should:
- know the structure of DNA
- know that chromosomes contain DNA
- know that genes are lengths of DNA and alleles are different forms of the same gene.

1.5 Monohybrid inheritance

Key terms

Genotype – Shows the alleles present in an individual, for a particular characteristic.

Phenotype – Visible characteristic of an individual.

Anyone working in health should understand inheritance. Horticulturists and agriculturists working in plant or animal breeding also need to understand inheritance patterns.

Monohybrid inheritance is the inheritance pattern for a characteristic governed by a *single* gene.

You have seen in lesson 1.4 that mutations can change the sequence of bases in a gene (part of a DNA molecule). The mutation gives rise to a different allele (version) of that gene.

Alleles for haemoglobin

Let's think about a characteristic such as the oxygen-carrying protein, **haemoglobin**, which is in your red blood cells.

- The instructions for making haemoglobin are in a gene on one of your chromosomes.
- You have two copies of the gene because you have 23 pairs of chromosomes in the nucleus of your cells. One member of each pair came from your mother and one came from your father.
- Most people have two normal alleles of the gene for making haemoglobin. They are homozygous for this characteristic. All their haemoglobin is normal. Normal haemoglobin is the **dominant** characteristic so these individuals are homozygous dominant.
- We can write their **genotype** for homozygous dominant as **HH**.
- Their **phenotype**, the visible characteristic, is that they have normal haemoglobin.

Sickle cell disease

Did you know?

Haemoglobin is a protein made of four separate chains of amino acids. Each chain has an iron atom attached to it and each iron atom can carry one molecule of oxygen.

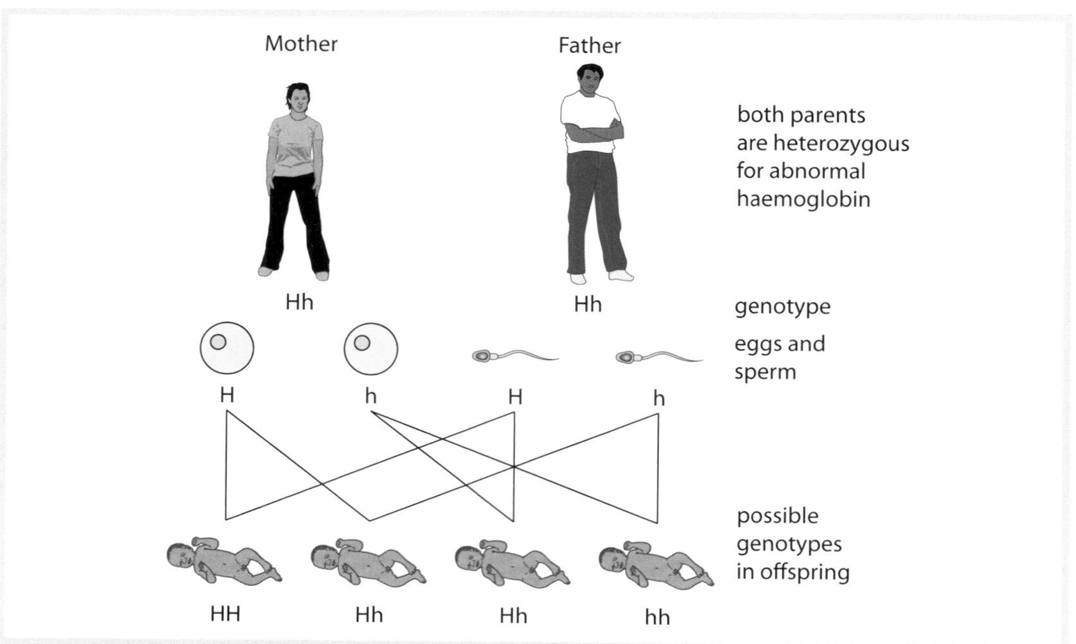

This genetic diagram shows that if both parents are heterozygous for sickle cell haemoglobin, the chance of a child having sickle cell disease is 1 in 4 or 25%.

If a mutation happened when sex cells (eggs and sperm) were being made in one parent, then some eggs or sperm could have a different allele for haemoglobin.

- Some of their offspring may inherit the new allele.
- However, they would also inherit the original (unmutated) version of the gene from the other parent.
- These particular offspring would be heterozygous for that characteristic.
- Their genotype would be written as **Hh**.
- The allele, **h**, codes for slightly different haemoglobin that does not have the usual shape. However, these people still have one normal allele so enough of their haemoglobin is normal for them to have no illness or symptoms. But they are carrying an allele for abnormal haemoglobin.

What happens if two carriers have children?

If two heterozygous people have children they might produce a child who has two abnormal alleles for haemoglobin.

- The child's genotype would be **hh** and all their haemoglobin would be abnormal.
- They are homozygous for the **recessive** characteristic, or homozygous recessive.
- This type of haemoglobin does not fit so well into red blood cells and can make the cells sickle shaped.
- These sickle-shaped red blood cells may not always be able to squeeze through the capillaries, so tissues would be starved of oxygen. This child would have **sickle cell disease**.

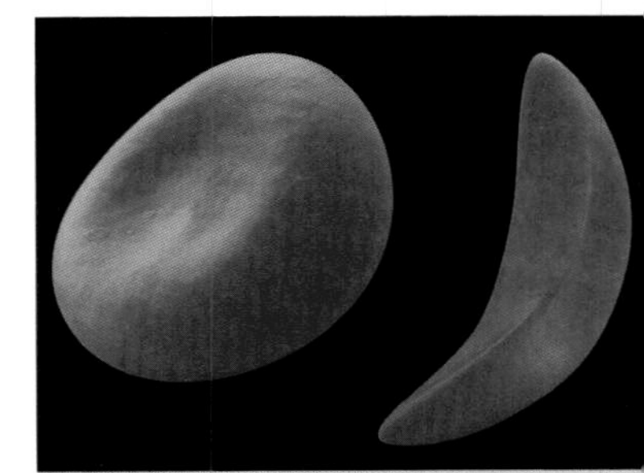

Normal red blood cell and sickled red blood cell.

Activity A

When body cells divide, the DNA in each of the chromosomes makes a copy of itself. Then the nucleus divides into two so that each new cell has a full set of chromosomes.

People with sickle cell disease need new red blood cells regularly because red blood cells cannot divide once they are in the blood stream.

Why do you think red blood cells cannot divide? (Hint: look back at lesson 1.2.)

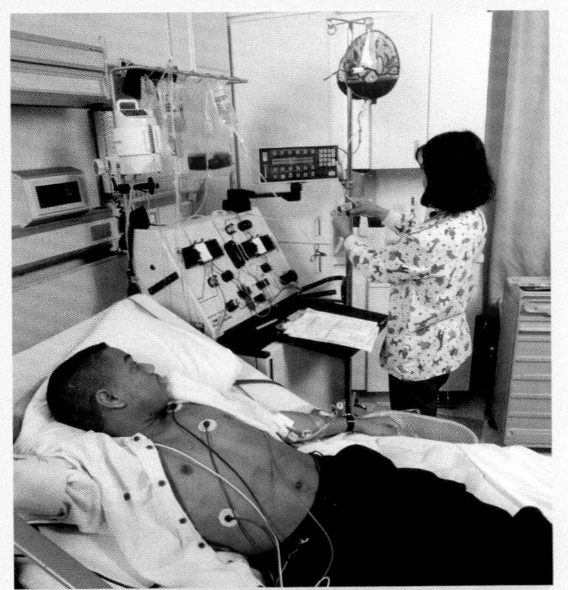

A patient with sickle cell disease needs to have their blood replaced through regular blood transfusions.

Lesson outcome

You should understand how monohybrid inheritance works.

1.6 Studying inheritance patterns

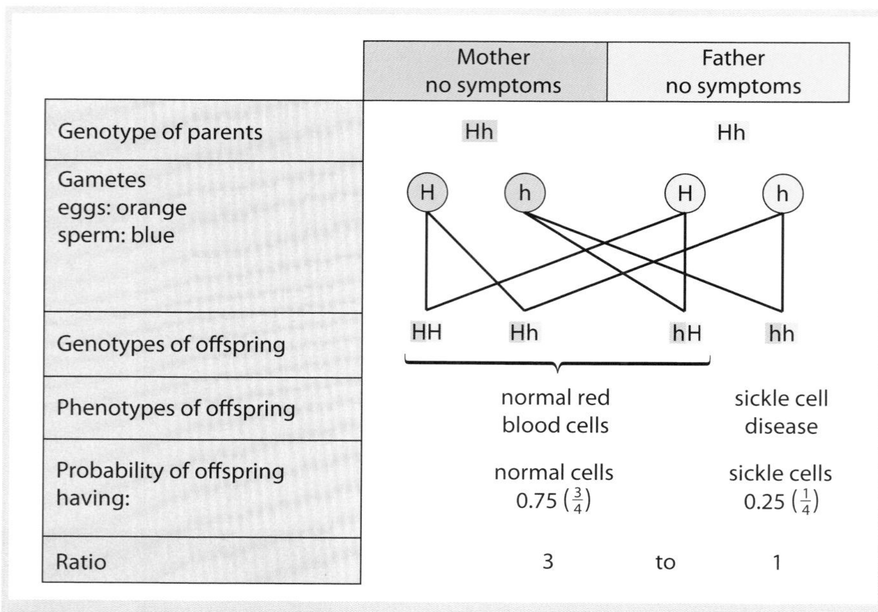

Genetic diagram showing the probability of a child of two heterozygous parents being born with sickle cell disease. The probability is the same for each pregnancy.

Genetic diagrams

We can use genetic diagrams to show the **probability** (chance) of a child being born with a characteristic inherited from two heterozygous parents.

Remember that eggs and sperm only have *one* allele for each characteristic.

Bear in mind as well that genetic diagrams show all genotypes which could be present when a sperm fertilises an egg.

Link

See lesson 1.5 for more about sickle cell disease.

Fair skin is a mutation that can be beneficial for people living in cooler climates.

Mutations can be harmful or beneficial

You can see from the example of sickle cell disease that to inherit two faulty alleles (hh) is very harmful. The blood of people who suffer from sickle cell disease cannot deliver enough oxygen to their respiring tissues. Their cells cannot function without energy to carry out their chemical reactions.

However, mutations can also be beneficial. Having fair skin is the result of mutations to genes controlling skin colour. It is an advantage to people living in cooler climates like the UK, as their skin can still make vitamin D from the less intense sunlight there.

Punnett squares

Punnett squares are another way of working out the probability of offspring inheriting a particular characteristic.

Father's genotype Hh / Mother's genotype Hh		Father's sperm genotypes	
		H	h
Mother's egg genotypes	H	HH	Hh
	h	Hh	hh

These four squares show possible genotypes of offspring

$\frac{3}{4}$ of the offspring have no symptoms

$\frac{1}{4}$ (25%) have sickle cell disease

Punnett square used to analyse monohybrid inheritance – both parents being heterozygous for sickle cell disease. What proportion of the offspring are homozygous dominant?

Pedigree analysis

A pedigree is a family tree diagram. It can be used to study the inheritance of a particular characteristic in a family.

- Females are shown as circles and males as squares.
- Those having the characteristic in question are shown with their circles or squares shaded.
- Individuals known to be carriers are shown as half shaded squares or circles.

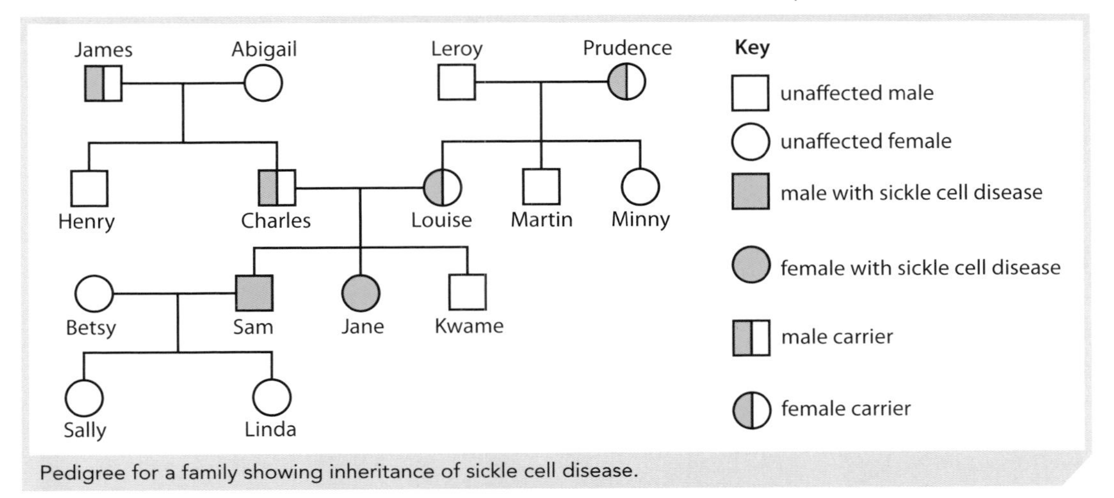

Pedigree for a family showing inheritance of sickle cell disease.

Activity A

Look at the pedigree analysis diagram.

1. Explain how Charles and Louise know they are carriers of sickle cell disease.
2. What are the genotypes of Sam and Jane?
3. Betsy has had a genetic test and knows that she is not a carrier of sickle cell disease. Draw a genetic diagram or a Punnett square to show the possible genotypes and phenotypes of Sally and Linda.
4. Prudence and James both know they are carriers of sickle cell disease although they have not had a genetic test. How do you think they know they are carriers?

Just checking

Remember that plants have genes too.

1. A type of plant may have red flowers or white flowers. Having white flowers is a recessive characteristic and having red flowers is a dominant characteristic. For each of the following genotypes, state its phenotype (visible characteristic): **(a)** RR, **(b)** Rr, **(c)** rr.
2. If a plant that is heterozygous for flower colour is crossed with a plant with white flowers, what colour flowers will the offspring have? Choose the correct answer from the following.
   - (a) 100% red flowers
   - (b) 100% white flowers
   - (c) 50% white flowers and 50% red flowers
   - (d) 25% white flowers and 75% red flowers
   - (e) 25% red flowers and 75% white flowers.

Lesson outcomes

You should:

- be able to work out the ratio of offspring inheriting particular characteristics using Punnett squares, genetic diagrams and pedigree analysis
- be aware that mutations can be beneficial or harmful.

1.7 Homeostasis

Key terms

Homeostasis – Keeping the body's internal conditions in a steady state.

Osmosis – Movement of water from a region where there are many water molecules (high water potential) to a region of lower water potential through a partially permeable membrane, usually a cell membrane.

Keeping things constant

Homeostasis is how living things keep their internal conditions constant. The table below shows three of the conditions that need to be kept in balance.

Internal condition of body	Why does it have to be maintained at a constant level?	Which organs and systems are involved in keeping it steady?
Temperature	If it was too cold your chemical reactions would be too slow. If it was too hot your **enzymes** would no longer work and your chemical reactions would not run. Your cells would also be damaged. Animals that can regulate their body temperature can live in a range of environments.	Nervous system Blood Skin Liver
pH	Your enzymes will not work properly if the pH is too high or too low. Your blood and cytoplasm pH can be just below or around pH 7. The only place in your body where the pH can be low is your stomach.	Nervous system Blood Kidneys
Blood glucose concentration	It has to be high enough for your cells to receive enough glucose for respiration to release energy. It should not be too high otherwise your blood cells, and other cells, will lose water by **osmosis**, shrivel up and stop working.	Endocrine system – cells in the **pancreas** Liver

Rock pigeons can regulate their body temperature. They are found everywhere in the world except the poles.

Many of the mechanisms in your body that are involved with homeostasis rely on something called **negative feedback**. This means as a change to the body happens in one direction, mechanisms in the body work to make it change in the opposite direction. This helps to keep conditions in the body balanced at around the right level.

You can see from the table above that the **nervous system** and the **endocrine** (hormone) **system** are both involved with homeostasis.

Activity A

Find out what a thermostat is and how it works. In small groups, discuss why this is an example of negative feedback.

The nervous system

The nervous system is made up of:

- the brain and spinal cord – together these are called the **central nervous system (CNS)**
- the **peripheral nervous system (PNS)** which is the **nerves**.

Nerves contain two types of neurones (nerve cells).

- **Sensory neurones** transmit electrical impulses from sensory organs to the CNS.
- **Motor neurones** transmit electrical impulses from the CNS to **effectors** (muscles and glands).

The nervous system communicates using mainly electrical impulses along neurones.

Human nervous system. The brain and spinal cord are the central nervous system. The nerves are the peripheral nervous system.

Discussion point

At lower temperatures conduction of nerve impulses slows down. Why do you think placing ice on your hands or feet makes them numb?

The endocrine system

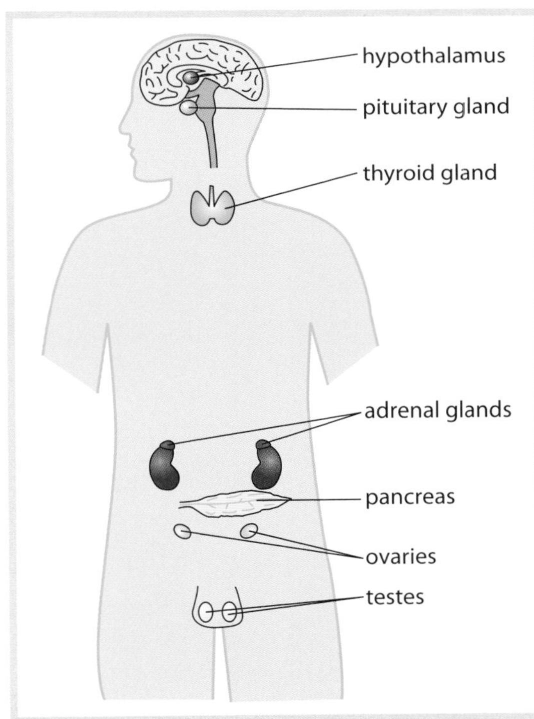

Diagram showing the positions of the main endocrine glands.

Endocrinologists are doctors who specialise in studying hormones.

The endocrine system consists of **glands** that release hormones straight into the blood stream. These hormones are chemical messengers. They cause certain parts of the body, called *target tissues* or *target organs*, to respond to their presence. They are carried all over the body by the blood, but only their target tissues or organs respond.

Did you know?

Too much growth hormone from the pituitary gland can make a person very tall. Too little means they do not grow enough. The world's tallest man was 2.72 m (8 ft 11 in). The current shortest man is 54.6 cm (1 ft 9.5 in).

Just checking

1. What does homeostasis mean?
2. Which two types of communication are involved in homeostasis?
3. The nervous system is made up of the CNS and the PNS. What are the CNS and PNS?

Lesson outcomes

You should:

- understand that homeostasis is the maintenance of a constant internal environment
- know that the nervous and endocrine systems are involved in homeostasis.

1.8 How nerves carry information

Key terms

Involuntary response – One not under conscious control; does not involve thought.

Voluntary response – One under conscious control; involves thought.

Diffusion – Movement of molecules from a region where they are in high concentration (lots of them) to a region where they are in lower concentration.

Link

Lesson 1.2 shows the structure of motor and sensory neurones.

Did you know?

Your brain can form a few million new connections every second. However, if you do not use your brain fully it will make fewer connections. It operates a 'use it or lose it' policy.

Neurones and synapses

Neurones are the longest cells in your body.

- They are specialised cells with long extensions for carrying electrical impulses.
- This transmission of information is extremely fast.
- Sometimes the information has to go across gaps between neurones. These gaps are called **synapses**.
- Electrical impulses cannot jump across these gaps. So, when an impulse arrives at the end of one neurone, it causes molecules of a chemical to be released.
- This chemical then moves across the gap by **diffusion** and joins onto special receptors on the cell membrane of the neurone on the other side of the synapse.
- When this happens it triggers electrical impulses to carry on along the second neurone.

Synapses – the gaps between the (swollen green) ends of the neurones.

Why do we have synapses?

Synapses allow neurones to communicate with lots of other neurones, making many nerve impulse pathways possible. This is very important, especially in the brain, which contains about 100 billion neurones. The more communication pathways, the more you can learn.

- Synapses also ensure that nerve impulses only travel one way, and so do not go backwards.
- Synapses may transmit some impulses and inhibit others, so they are essential for homeostasis.

Involuntary and voluntary responses

- A reflex is a simple behaviour pattern that allows a fast response to a stimulus.
- Most reflexes protect us from harm and so they need to be quick and automatic.
- You do not think about them or decide whether or not to respond, so they are also described as **involuntary**.

When you were a baby you had primitive reflexes, such as grasping and suckling. These reflexes ensure that babies make contact with the mother and can feed. You have lost these reflexes but you can still jump up if you sit on a pin, blink if something is moving towards your eyes and cough or gag if something is tickling your throat.

The knee jerk reflex

Your knee jerk reflex isn't there just so the doctor can test your reflexes. You use it all the time when standing or walking to keep you upright.

- As you stand your knees tend to buckle under the weight of your body.
- As soon as your knees start to bend your thigh muscles are stretched.
- This stretching acts as a **stimulus** to the stretch **receptors** in the muscle.
- These receptors transmit impulses along sensory neurones to the spinal cord.
- In the spinal cord, the sensory neurones connect, via synapses, to motor neurones.
- Motor neurones then send impulses back to the thigh muscle (the effector) and cause it to contract.
- This pulls the lower leg up and straight.

This reflex is happening all the time as you stand and walk, without you realising it. It is very fast. It happens about 20 times per second.

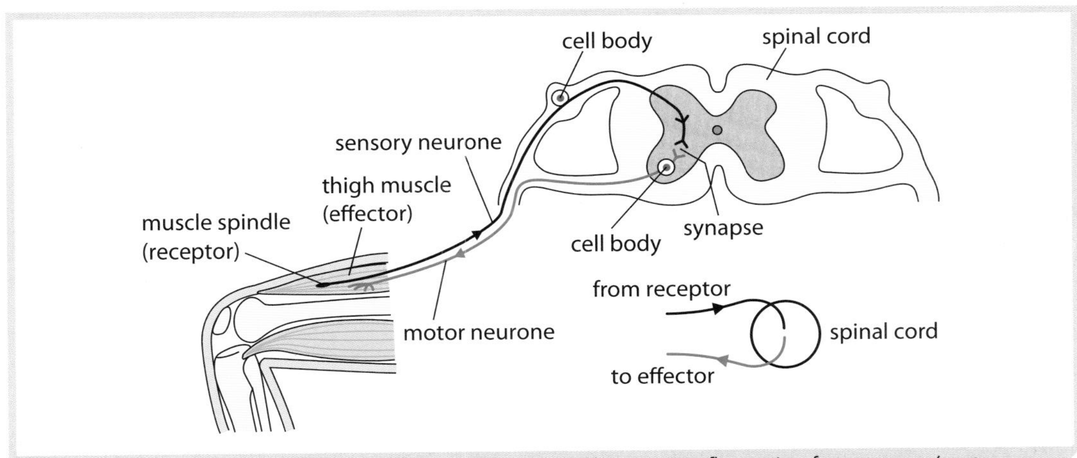

The nerve pathway involved in the knee jerk reflex. The pathway in a reflex action forms an arc (part of a circle) shape, so it is called the **reflex arc**.

Voluntary responses

Voluntary responses involve thought and are under the conscious control of your brain. For example, raising your hand in class is a voluntary response.

Just checking

1. What is the difference between voluntary and involuntary responses?
2. If you stroke the cheek of a newborn baby it roots – turns its head and begins to search for food. What type of response is this?
3. Describe the reflex arc involved when you step on a drawing pin with a bare foot.

Discussion point

Sometimes you can override a reflex or learn to ignore it. People wearing contact lenses have to overcome the blinking reflex. Can you think of any other examples of overriding a reflex?

Activity A

Nerve impulses travel along neurones very fast. There are variations, but they can reach speeds of up to 130 metres per second.

1. Calculate how long a nerve impulse takes to get from your big toe to your spinal cord in the lower region of your back. Assume that the distance it travels is 1.3 m.
2. Estimate how long it takes between stubbing your toe and moving your foot away.
3. The response described here is an involuntary reflex. However, you are aware of stubbing your toe because you feel pain. Pain is a perception and that involves your brain. How do you think the brain 'knows' you have stubbed your toe?

Lesson outcomes

You should:

- know the difference between involuntary and voluntary responses
- know the structure of a simple reflex arc and how it protects the body from harm.

1.9 Two examples of homeostasis

Key term

Glycogen – Type of carbohydrate, sometimes called 'animal starch'. It is a large molecule made of many glucose molecules joined together.

1 Using hormones

Your blood glucose concentration is regulated through negative feedback. Remember this just means that as your blood glucose level changes, your body responds to bring the level back to normal. The hormones involved are **insulin** and **glucagon**.

Some nurses specialise in running clinics to help **diabetics** live as normally as possible. These nurses need to fully understand how insulin and glucagon work.

The stimulus

The stimulus is the *change* in blood glucose concentration.

- After you have eaten a meal, your blood glucose concentration will *go up* as more glucose is absorbed from the small intestine into your bloodstream (point **1** in diagram).
- If you miss a meal, your blood glucose will *fall* as your cells have used it for their respiration (point **4**).

Did you know?

What a pregnant woman eats can affect the health of her baby for the rest of their life. If she doesn't eat enough protein while pregnant, some of the baby's organs may not develop properly.

While the baby may appear physically 'normal', they may not have enough cells to make insulin in their pancreas. This may increase their chance of developing type 2 **diabetes** later in life.

Receptor and effector

- In each case, the receptor is cells in your pancreas. Chemical sensors inside these cells detect the change in blood glucose concentration.
- Cells in your pancreas are also the effector as they respond by producing hormones.

The hormones

- When your blood glucose level *rises*, the cells in your pancreas secrete the hormone, insulin (point **2** in diagram).
- When your blood glucose level *falls*, the cells in your pancreas secrete the hormone, glucagon (point **5**).

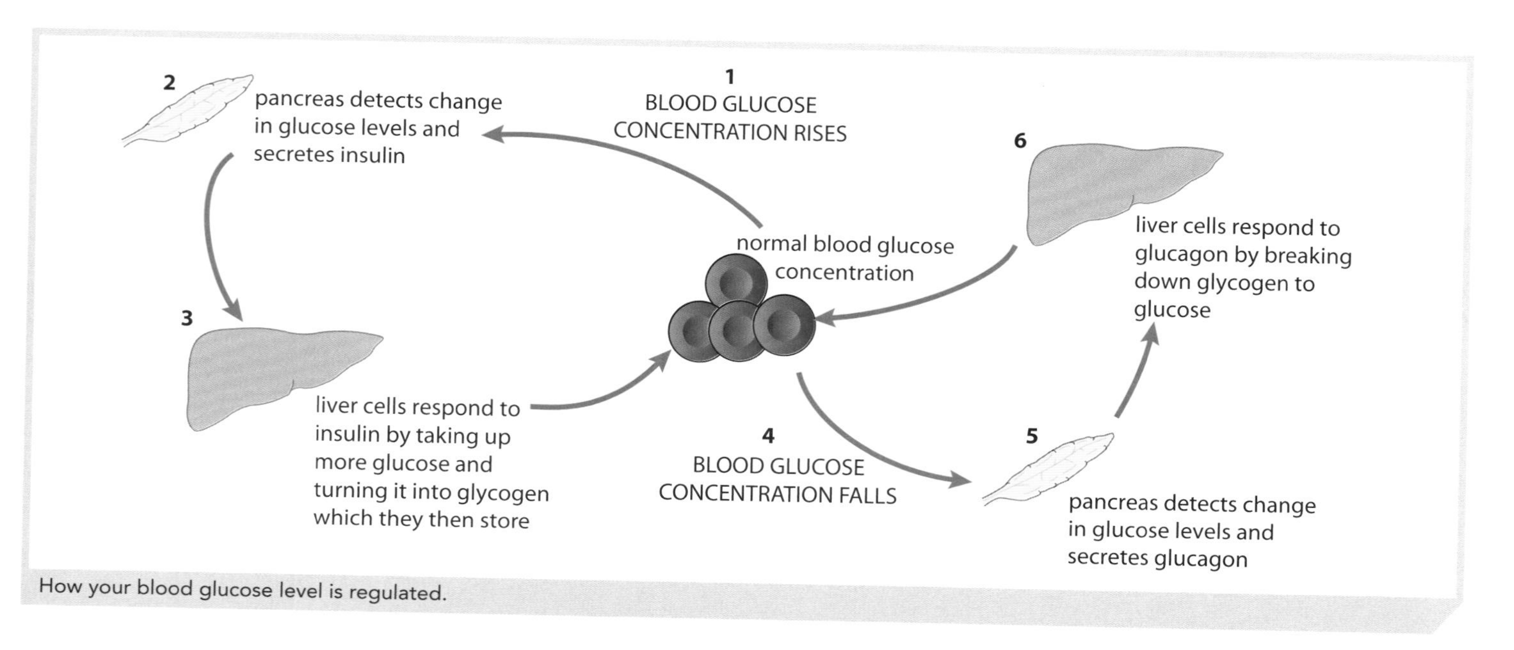

How your blood glucose level is regulated.

The target organ

Both insulin and glucagon are carried in your blood to your **liver**, which is the target organ.

- **Insulin** causes your liver cells to take up more glucose from your blood and turn it into **glycogen** for storage. This lowers the blood glucose level, bringing it back to normal (point **3** in diagram).
- **Glucagon** causes your liver cells to break down glycogen to glucose and release this glucose into your blood, bringing the level back up to normal (point **6**).

2 Using the nervous system

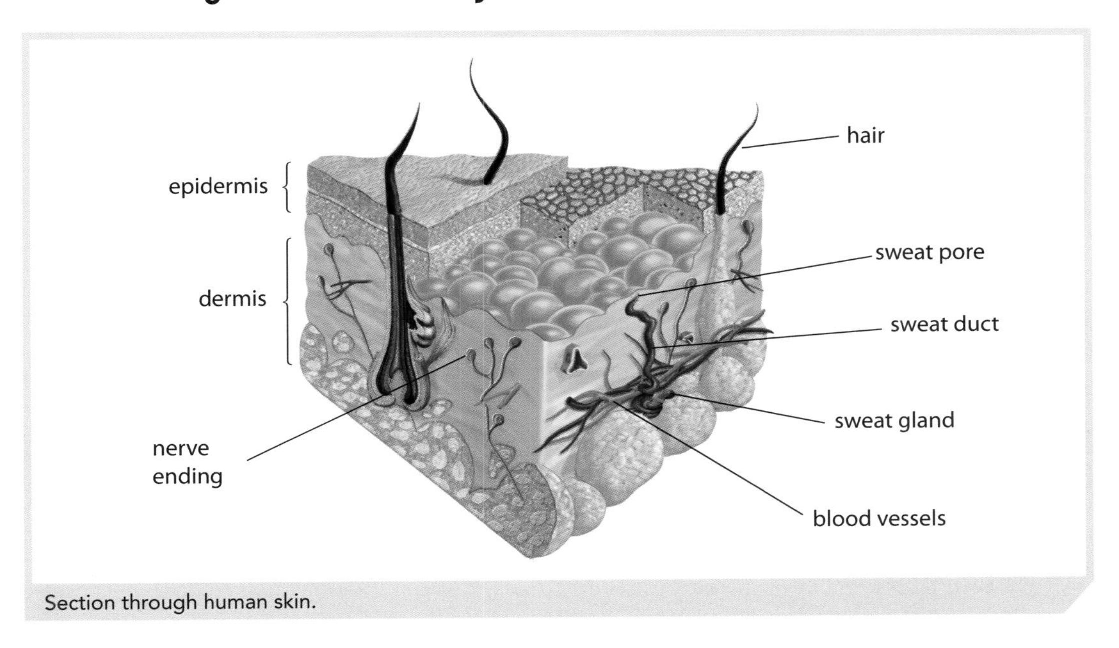

Section through human skin.

Did you know?

Your skin is your largest organ. The skin of an average-sized person covers an area of about 2 m^2 and weighs about 5 kg.

Activity A

Besides the responses your body makes to maintain your temperature, you can also alter your behaviour to help you cool down or keep warm. Think of some of these behavioural ways of maintaining temperature. Write a list and then compare your thoughts with those of others in your class.

Your body temperature is regulated using the nervous system. When you are hotter or colder than usual, receptors in your skin detect the change and send impulses to a part of your brain. This part of the brain then sends impulses back to your skin and causes it to respond. Examples of how the skin responds are shown below and over the page.

When you are hot

- Increased sweating – sweat glands release more sweat onto your skin. Heat from your body evaporates water in the sweat. This cools you.
- **Vasodilation** – blood vessels in the skin dilate (widen) so more blood flows near the surface of your skin. The blood loses heat by radiation.
- Hair erector muscles relax and hairs lie flat so heat can be lost more easily from your skin by convection.

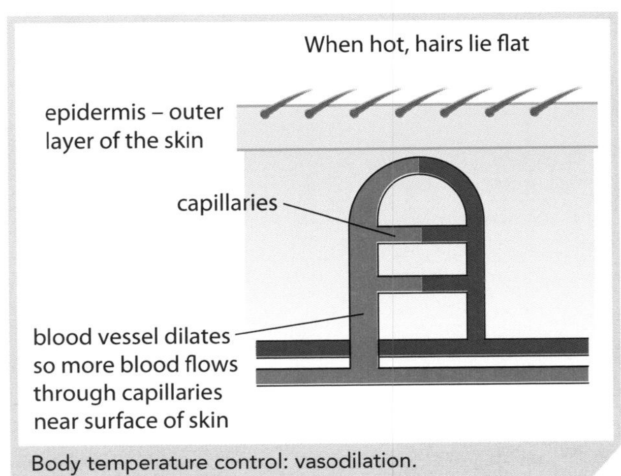

Body temperature control: vasodilation.

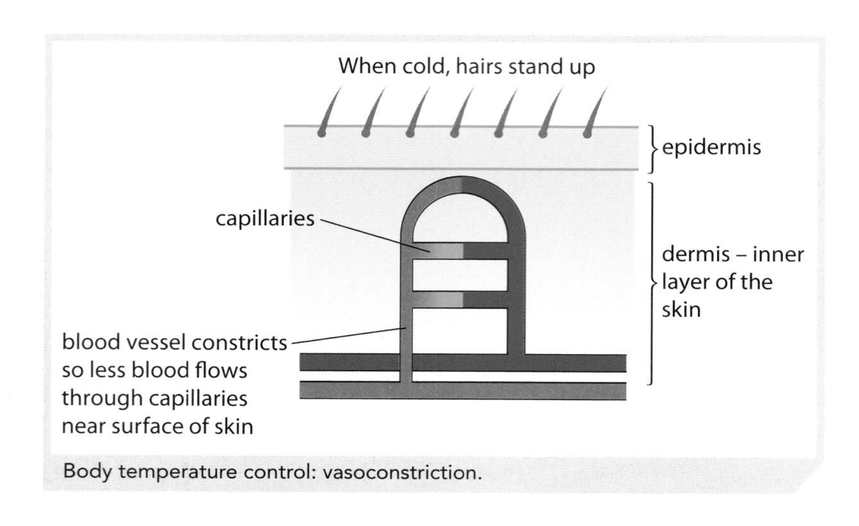

Body temperature control: vasoconstriction.

When you are cold

- Decreased sweating – so heat is not lost by evaporation.
- **Vasoconstriction** – blood vessels in the skin constrict (become narrower) so less blood flows through them and less heat is lost by radiation.
- Shivering – muscles contract and expand very quickly. This shivering releases heat which warms you up.
- Hair erector muscles contract and hairs stand up. This traps air and helps insulate us. In the case of humans it does not help much, but you can see goose pimples where the hair muscles are contracting. In animals with fur or feathers, this response greatly helps them keep warm.

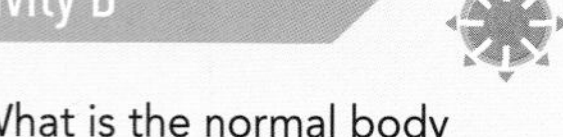

Activity B

1. What is the normal body temperature for humans?
2. Describe the three ways that your body loses heat and three ways that it prevents heat loss and helps keep you warm.

Differences between nervous and hormonal communication

Now that you have read about some examples of homeostasis, you can see that there are differences in the way the nervous system and the endocine system communicate with the body.

Feature	Nervous system	Endocrine system
Speed of communication	Very fast	Slower
Method of transport or transmission	Electrical impulses along neurones with chemicals at synapses between neurones.	Chemicals called hormones which are carried in the blood to the target organs.
Duration (length) of response	Short-term	Usually longer-term. Many are involved with regulating growth and development.

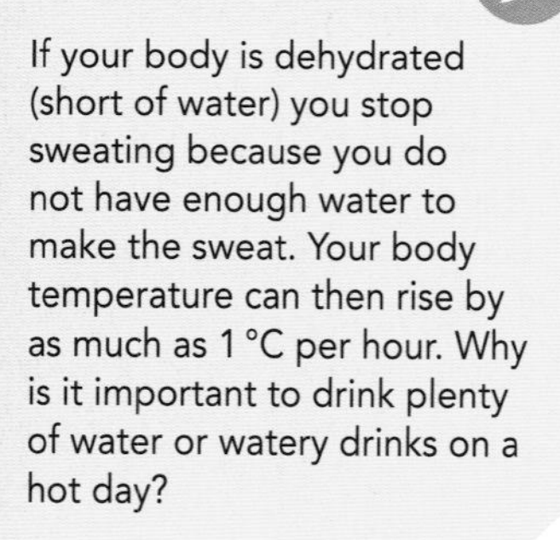

Discussion point

If your body is dehydrated (short of water) you stop sweating because you do not have enough water to make the sweat. Your body temperature can then rise by as much as 1 °C per hour. Why is it important to drink plenty of water or watery drinks on a hot day?

Lesson outcomes

You should:

- understand the roles of the hormones insulin and glucagon
- know the mechanisms involved in body temperature regulation
- appreciate some of the differences between nervous and hormonal communication.

Just checking

1. When you are cold your body responds by shivering and the hairs on your skin stand up. You also put on more clothes. Which of these responses is/are involuntary and which is/are voluntary?
2. How are hormones carried to their target organs?
3. How is the structure of neurones adapted to enable them to carry out their function?
4. Which system regulates your growth and development by producing chemicals into your bloodstream?

WorkSpace

Hannah Verghese

Advocacy and Policy Manager

The Migraine Trust

This is the UK's leading health and medical research charity for migraine and was established in the early 1960s. We support and inform people who suffer with migraine, and also help to educate health professionals and raise funding for research in this area.

More people suffer with migraine than asthma, diabetes and epilepsy combined. There are approximately 190 000 migraine attacks each day in the UK. So, this is a serious public health issue, but many people are not aware of the scale of the condition and how debilitating it can be for those who have it. It is often undiagnosed and widely under-treated, for both adults and children.

It is the job of The Migraine Trust to raise awareness in the public, media and Parliament.

I run the advocacy service that supports people with migraine to understand their rights and ensure that they get treated fairly at work, in school and by the NHS. If a migraine sufferer is being treated unfairly I may act on their behalf to try to change the situation. I also run workshops and speak at events for migraine sufferers and health professionals.

Recently, I was involved in raising awareness, amongst teachers and school nurses, of the problems when children suffer migraine attacks and cannot always access their medication in time. Many adults working with such children are unaware of the problem and this greatly adds to the distress of the children.

We work with other charities to ensure that migraine is part of the national health agenda.

The Migraine Trust helps to fund research projects in many parts of the world to try to find the cause of migraine. These projects:

- look at the biology of the condition
- research areas of the brain active during migraine attacks, and the differences between people with and those without migraine
- determine how to improve the management and treatment of migraine.

Migraine has a genetic and an environmental component, and the latest research indicates that it involves dysfunction of sensory nervous processing in the brain stem. As well as an intense headache, sufferers experience nausea and sensitivity to light and smells. Before the onset they may feel very tired and irritable and cannot concentrate.

Think about it

1. Why is it important that teachers and nurses in schools and colleges are better informed about migraine?
2. Why is research into the cause of the disease important?

1.10 Atomic structure

Get started

All matter in the Universe is made up of atoms. But how much do we know about atoms? Discuss with a partner whether these statements are true or false:

- Atoms cannot be split up into anything smaller.
- Atoms are solid objects and nothing can pass through them.
- There are millions of different kinds of atoms.
- No-one has ever seen an atom.

What is an atom made of?

All matter is made up of particles called **atoms**. Atoms are very, very small. It is not possible to see an atom even under the most powerful light microscope. However, scientists and engineers have developed ways of producing images of atoms using other methods.

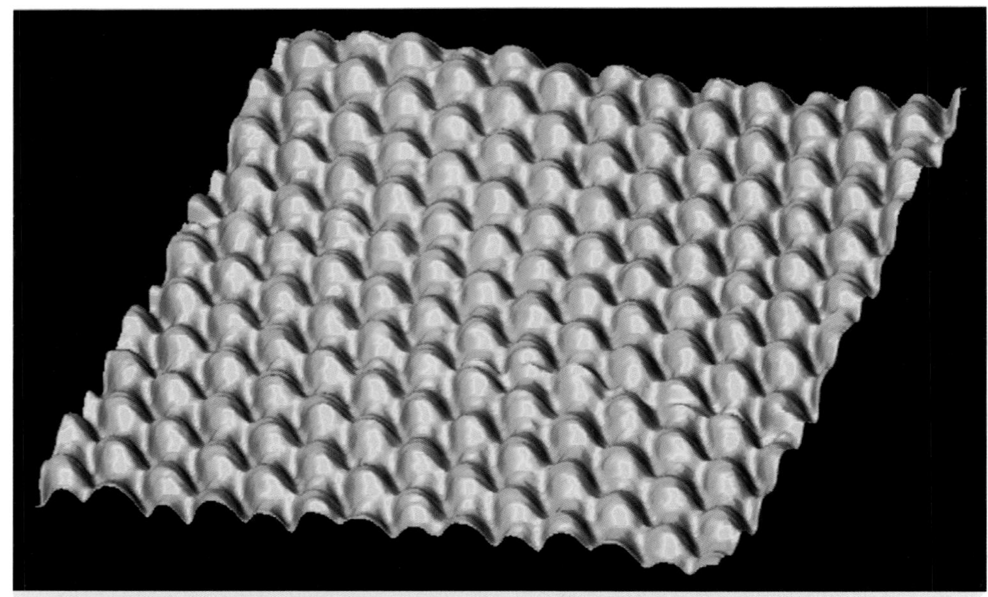

An image showing atoms on the surface of a metal, produced by a method called scanning tunnelling microscopy.

About 200 years ago, when scientists first realised that all matter was made up of atoms, they thought that atoms were the smallest particles. But now we know that atoms are made up of even simpler sub-atomic particles – **protons**, **neutrons** and **electrons**.

At the centre of an atom is a positive nucleus. If an atom were the size of The O2 arena, the nucleus would be the size of a ping-pong ball.

The nucleus is surrounded by electrons, arranged into layers or shells. Sometimes these **electron shells** are called **energy levels**.

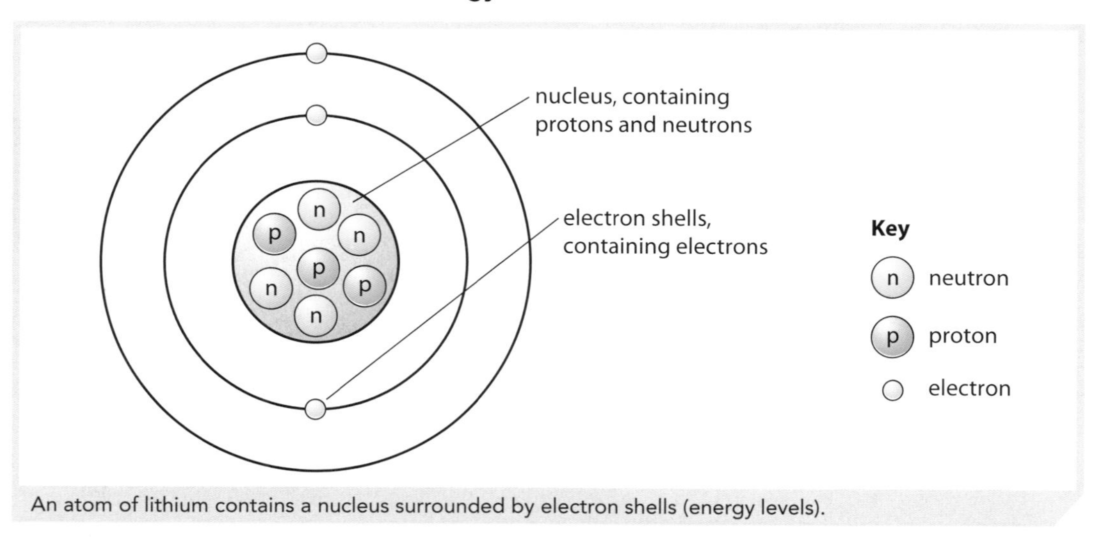

An atom of lithium contains a nucleus surrounded by electron shells (energy levels).

Key terms

Element – A substance which contains just one type of atom.

Atomic number – The number of protons in the nucleus of an atom.

Mass number – The total number of protons and neutrons in the nucleus of an atom.

Did you know?

Because the nucleus is so small, most of the atom is actually empty space.

When scientists tried firing sub-atomic particles at a thin gold film, they found that most of them passed straight through.

Only a tiny number of the sub-atomic particles hit the nuclei of the gold atoms and bounced back.

The sub-atomic particles

Each of the sub-atomic particles has different properties, as shown in this table.

Particle	Relative charge	Relative mass	Where in the atom is it found?
Proton	+1	1	In the nucleus
Neutron	0	1	In the nucleus
Electron	−1	Almost 0 (actually 0.0005)	In shells surrounding the nucleus

Link

Lesson 1.12 discusses the arrangement of elements in the periodic table.

Atomic number and mass number

So far scientists have identified 118 different types of atom. This means there are 118 different **elements**.

Each element has a fixed number of protons in the nucleus. For example, all atoms of lithium have 3 protons in the nucleus. This is the **atomic number** for lithium. There are no other elements with an atomic number of 3.

The **mass number** of an atom is the total number of protons and neutrons in the nucleus. Lithium has 4 neutrons, so its mass number is 7. The **nuclear symbol** for an atom is its chemical symbol with the atomic number and mass number.

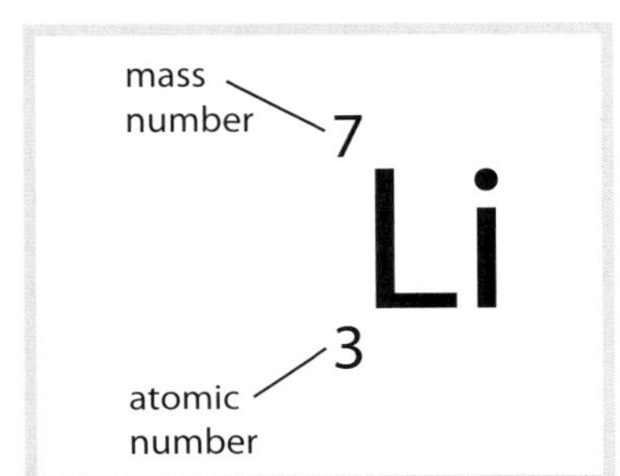

The nuclear symbol for the lithium atom shows you the atomic number and mass number of the atom.

Activity A

1 Using the nuclear symbols shown for these atoms, write down the numbers of protons and neutrons in the nucleus of each atom.

$^{16}_{8}O$ $\quad$ $^{23}_{11}Na$

2 An atom of potassium (symbol K) has 19 protons and 20 neutrons. Write down the nuclear symbol for this atom.

Worked example

You can use the nuclear symbol for an atom to work out the number of protons and neutrons in the nucleus. If a phosphorus atom has a nuclear symbol $^{31}_{15}P$, how many protons and neutrons are there in its nucleus?

Step 1 The atomic number is 15. This is the number of protons in the nucleus.

Step 2 The mass number is 31. This is the total number of protons and neutrons in the nucleus.

Step 3 So, the number of neutrons = 31 − 15 = 16.

Take it further

Neutrons have an important role to play in the nucleus of an atom. They prevent the positive protons from repelling each other and blowing the nucleus apart.

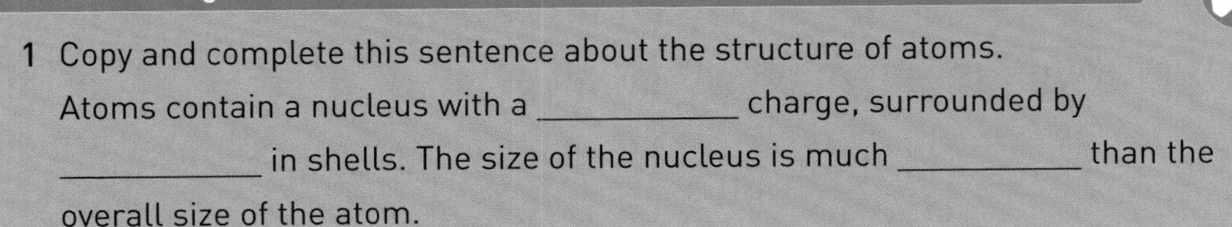

Just checking

1 Copy and complete this sentence about the structure of atoms.

Atoms contain a nucleus with a ____________ charge, surrounded by ____________ in shells. The size of the nucleus is much ____________ than the overall size of the atom.

Lesson outcomes

You should know:

- about the structure of an atom
- about the relative size of the nucleus
- that atoms of a given element have the same number of protons in the nucleus
- the meaning of the terms 'atomic number' and 'mass number'.

1.11 Isotopes and relative atomic mass

Key terms

Isotopes – Atoms of the same element which have the same number of protons but different numbers of neutrons.

Relative atomic mass (of an element) – The average mass of an atom of an element compared to a standard mass.

Two kinds of carbon atoms?

Almost all of the carbon atoms found in living things and in the rocks, seas and atmosphere of our planet have the nuclear symbol $^{12}_{6}C$. This is called carbon-12. It is so stable that the nuclei of these carbon atoms have not changed since the **Solar System** was formed over 4 billion years ago.

A tiny fraction of the carbon atoms on Earth have the nuclear symbol $^{14}_{6}C$. This is known as carbon-14. The nuclei of carbon-14 atoms have two extra neutrons when compared with carbon-12, so the two atoms have different mass numbers.

The two different carbon atoms are called **isotopes** of the element carbon. There is also a third isotope of carbon with a mass number of 13. It has the symbol $^{13}_{6}C$.

Link

Lesson 1.10 explains how to work out the number of protons and neutrons from the nuclear symbol.

Did you know?

Carbon-14 is **radioactive**. This means that when a living organism dies, the carbon-14 atoms it contains gradually decay into more stable atoms. Scientists can use this fact to find the age of the remains of long-dead organisms. This is called carbon dating.

Ötzi the 'Iceman' was found in a glacier in the Alps. Archaeologists used the amount of carbon-14 remaining in his body to calculate that he died about 5000 years ago.

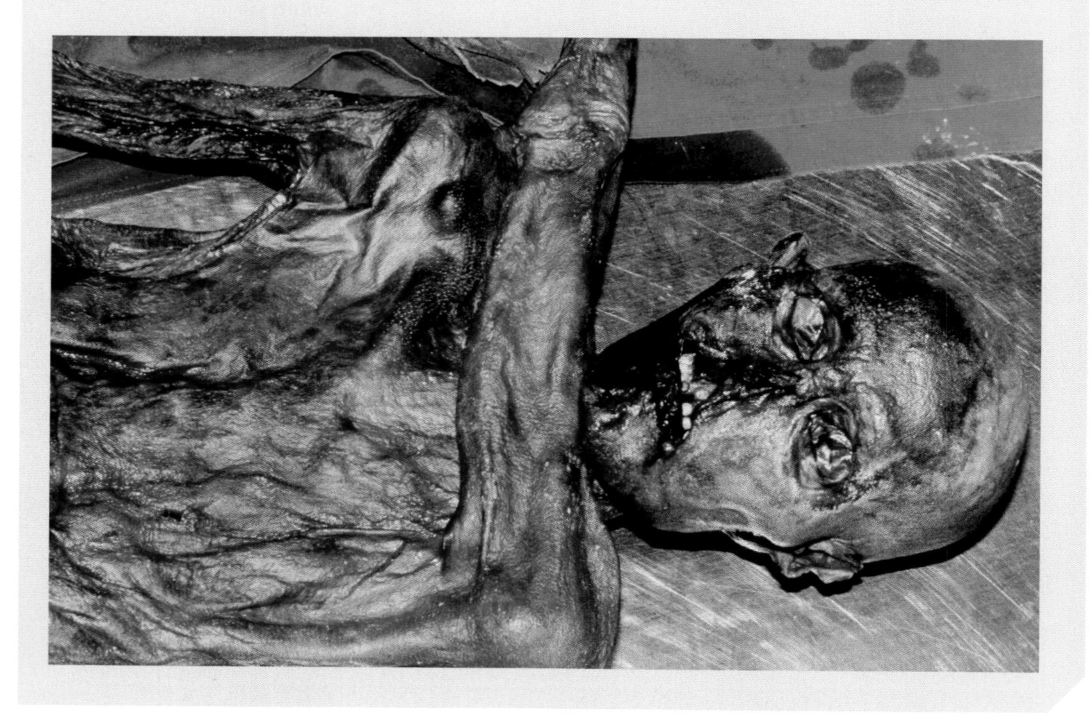

Activity A

Write down the number of protons and neutrons in the following atoms:

- carbon-12
- carbon-14.

Take it further

Relative atomic mass is a way of comparing the mass of an atom to a standard mass. In the past, all masses were compared to a hydrogen atom – but now they are all compared to 1/12 of the mass of a carbon-12 atom.

Relative atomic mass

Only a tiny fraction of carbon atoms have a mass number of 13, or of 14. All the rest have a mass number of 12. So, the average mass of a carbon atom is 12.01. This is called the **relative atomic mass** of carbon.

Isotopes and relative atomic mass

The most common isotope of chlorine is $^{35}_{17}Cl$. But about a quarter of all chlorine atoms are a different isotope: $^{37}_{17}Cl$. Because this is quite a high proportion, the relative atomic mass is not a whole number – it actually works out to be 35.5.

Assessment tips

You may be required to use data about the abundance of isotopes to predict which elements will have relative atomic masses that are not whole numbers.

You also need to be able to calculate the relative atomic mass from these data.

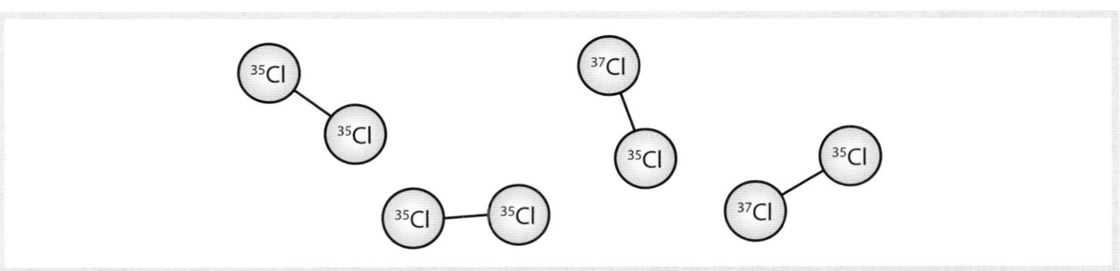

About three-quarters of all chlorine atoms are chlorine-35 isotopes. The rest are chlorine-37.

There are many other elements like chlorine, where the relative atomic mass is not a whole number because of the presence of different isotopes. The table shows three examples.

Element	Relative atomic mass	Isotopes present and percentage abundance
Lithium	6.9	$^{6}_{3}Li$ (7.6%) $^{7}_{3}Li$ (92.4%)
Magnesium	24.3	$^{24}_{12}Mg$ (79%) $^{25}_{12}Mg$ (10%) $^{26}_{12}Mg$ (11%)
Copper	63.5	$^{63}_{29}Cu$ (70%) $^{65}_{29}Cu$ (30%)

Worked example

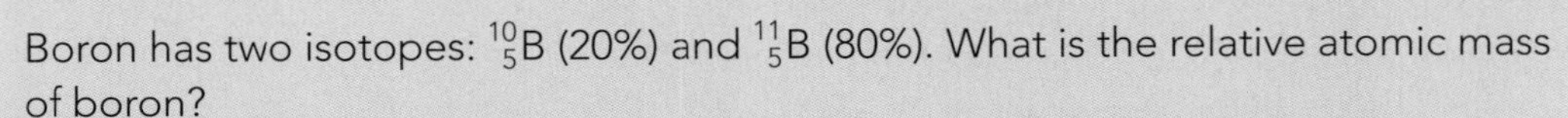

Boron has two isotopes: $^{10}_{5}B$ (20%) and $^{11}_{5}B$ (80%). What is the relative atomic mass of boron?

Step 1 First, multiply the mass number of each isotope by its percentage abundance.

Step 2 Then, add all these values together.

Step 3 Finally, divide by 100 to get the average.

So, for boron, the relative atomic mass $= \frac{(10 \times 20) + (11 \times 80)}{100} = \frac{1080}{100} = 10.8$

Activity B

The element thallium has two isotopes:

$^{203}_{81}Tl$ (30%) $^{205}_{81}Tl$ (70%)

Use this information to calculate the relative atomic mass of thallium.

Just checking

1 Look at these nuclear symbols for four atoms, A–D:

$^{14}_{7}A$ $^{16}_{8}B$ $^{15}_{7}C$ $^{14}_{6}D$

(a) How many of these atoms have eight neutrons in their nucleus?

(b) Which two of these atoms are isotopes of each other? Explain your answer.

2 Bromine has two isotopes: 50% of bromine atoms have a mass number of 81 and 50% of bromine atoms have a mass number of 79. What is the relative atomic mass of bromine? Explain your answer.

Lesson outcomes

You should know:

- about relative atomic mass and isotopes
- how to calculate relative atomic mass from percentage abundances.

1.12 The periodic table

Key term

Periodic table – A way of arranging elements in order of their atomic number to show the patterns in their properties.

Link

Lesson 1.10 discusses the useful information about elements that is displayed in the periodic table.

Take it further

Not all the elements in the periodic table belong to groups 1–0. If you look at the large periodic table at the back of this book, you will see that there is a block of elements between groups 2 and 3. These are all metals and are called the transition elements.

Patterns in science

There are patterns in the world around us. Sometimes the patterns are very obvious – like the way the Moon goes through phases in a repeating pattern. Sometimes you need to look harder to see the patterns – like the way living things inherit characteristics from their parents.

There are patterns in the ways chemical substances behave. These patterns are key to how the elements are arranged in the **periodic table**.

The periodic table

In the 19th century, scientists began to arrange elements in order of their relative atomic mass. When they did this, they noticed that there were patterns in the way in which the elements behaved. To help show this, scientists drew up a table. They found that elements with similar properties were grouped together.

Nowadays, we arrange the elements in order of their atomic number and the patterns show up even more clearly.

Groups (vertical columns)	Elements in the same group have similar properties.
Periods (horizontal rows)	Elements in the same period do not have similar properties, but the properties often change in a regular pattern as you go across.

1	2	3	4	5	6	7	0
1 H hydrogen 1							4 He helium 2
7 Li lithium 3	9 Be beryllium 4	11 B boron 5	12 C carbon 6	14 N nitrogen 7	16 O oxygen 8	19 F fluorine 9	20 Ne neon 10
23 Na sodium 11	24 Mg magnesium 12	27 Al aluminium 13	28 Si silicon 14	31 P phosphorus 15	32 S sulfur 16	35.5 Cl chlorine 17	40 Ar argon 18
39 K potassium 19	40 Ca calcium 20						

The columns are called groups.

metals

non-metals

The rows are called periods.

A short-form version of the periodic table. Metals and non-metals are shown in different colours. All relative atomic masses except chlorine are rounded to the nearest whole number. A larger version of the periodic table is shown at the back of this book.

Activity A

The element chlorine (Cl) has an atomic number of 17. Work in pairs and use the periodic table to answer the following questions about this element.

1 What group and what period of the periodic table is chlorine in?

2 Is chlorine a metal or a non-metal?

3 What are the other elements in the same group?

4 What can you say about the properties of the other elements in the same period?

Did you know?

The periodic table is expanding. About 98 elements occur naturally, but particle physicists can make new elements by making atoms travel very fast and smashing them together. This turns the smaller atoms into bigger atoms. The problem is that these new elements barely exist for even a fraction of a second because they are very unstable. They break down into smaller atoms again.

When this book was first published, elements up to 118 had been officially reported. You could search the Internet to see whether more have been reported since then.

Part of the Large Hadron Collider (LHC) in Switzerland. New elements have been created in particle accelerators like this one.

Metals and non-metals

Metals are found in the middle and on the left of the periodic table. Non-metals are found on the right of the periodic table.

Just checking

1 This question is about the atoms aluminium (Al), oxygen (O), sodium (Na) and sulfur (S).

   (a) Which two elements are non-metals?

   (b) Which two elements are in the same period?

   (c) A learner says that sulfur and oxygen might have similar properties. Explain how they can tell this from the periodic table.

Lesson outcomes

You should know

- how elements are arranged in the periodic table
- the position of metals and non-metals
- the fact that elements with similar properties are placed in the same vertical columns (called groups).

1.13 Electronic configurations

Key term

Electronic configuration – The way electrons are arranged in the shells of an atom.

Electrons surround the nucleus of an atom. The electrons are arranged in shells at different distances away from the nucleus. Electrons in each shell have different amounts of energy so the shells are sometimes called energy levels.

Scientists can predict how many electrons there will be in each shell of any atom. This is called the **electronic configuration** of an atom.

Link 

Lesson 1.10 explains how to work out the number of protons from the nuclear symbol.

Number of electrons in an atom

The number of electrons in an atom is the same as the number of protons. Protons have a charge of +1 and electrons of −1. So, if there are equal numbers of protons and electrons, the atom will be neutral (uncharged).

Rules for filling up shells

There is a limit to how many electrons can fit into a shell before it is full. Electrons fill up the shells starting with the lowest shell (closest to the nucleus). When one shell is full, the electrons must go into the next shell as follows. The following rules can be used for elements 1 to 20:

- first shell: a maximum of 2 electrons
- second shell: up to 8 electrons
- third shell: up to 8 electrons
- fourth shell: any more electrons

The electronic configuration of an atom can be shown in a diagram like the one for sodium (right). The configuration can be written as: 2,8,1. The first number tells you how many electrons there are in the first shell, the second tells you how many electrons there are in the second shell and so on.

eight electrons can fit into the second shell

the final electron goes into the third shell

Na

two electrons can fit into the first shell

The electronic configuration of a sodium atom.

Activity A 

Find the elements neon and potassium in the table of electronic configurations.

1 Write out the electronic configuration of a potassium atom (atomic number 19).

2 Draw out the electronic configuration of a neon atom (atomic number 10).

Worked example 

The atomic number of magnesium is 12. Predict the electronic configuration of an atom of magnesium and write it down.

Step 1 There are 12 electrons in a Mg atom.
Step 2 Two electrons can go into the first shell.
Step 3 Eight more can go into the second shell.
Step 4 This leaves two more electrons, which must go into the third shell.
Step 5 So the electron configuration is written as 2,8,2.

The table shows the electronic configuration of some of the first 20 elements in the periodic table.

Atomic number	Element	1st shell	2nd shell	3rd shell	4th shell
1	Hydrogen	1			
2	Helium	2			
3	Lithium	2	1		
4	Beryllium	2	2		
5	Boron	2	3		
6	Carbon	2	4		
7	Nitrogen	2	5		
8	Oxygen	2	6		
9	Fluorine	2	7		
10	Neon	2	8		
11	Sodium	2	8	1	
18	Argon	2	8	8	
19	Potassium	2	8	8	1
20	Calcium	2	8	8	2

The importance of the outer shell

It is very important to know how many electrons there are in the outer shell of an atom. The outer shell electrons are the ones involved in chemical reactions, so knowing how many electrons there are in this shell helps you to predict how an element will react.

Electronic configurations and the periodic table

There is a connection between the electronic configuration and the periodic table. The number of electrons in the outer shell is equal to the group number. For example:

- bromine has seven electrons in its outer shell, which means that it is in group 7
- all elements in group 1 have just one electron in the outer shell.

Elements in the same group have the same number of electrons in their outer shell, so they have similar properties.

Just checking

1 Fluorine has an atomic number of 9.
 (a) How many electrons are there in a fluorine atom?
 (b) Write down the electronic configuration of a fluorine atom.

2 An element X has electronic configuration 2,8,8,1.
 (a) What group is this element in?
 (b) Name one other element which will have similar properties to this element.
 (c) Draw a diagram to show how the electrons are arranged.

Activity B

The element barium has two electrons in its outer shell.

1 What group of the periodic table does it belong to?

2 Name another element which will have similar properties to barium.

Sodium (top) and potassium (bottom) both have one electron in their outer shell and react in a similar way with water.

Lesson outcomes

You should know:

- that atoms contain equal numbers of protons and electrons
- the rules for filling electron shells for the first 20 elements of the periodic table
- the connection between the number of outer electrons and the position of the element in the periodic table.

1.14 Elements, compounds and mixtures

Key terms

Element – A substance which contains just one type of atom.

Compound – A substance made up of two or more different elements chemically bonded together.

Molecule – A single particle made up of two or more atoms bonded together.

Mixture – A substance made up of two or more simpler substances that are not chemically bonded. Mixtures can contain elements and compounds.

Elements

Copper is an example of an **element**. It is represented by the symbol Cu. You can identify elements by remembering that:

- elements contain just one type of atom
- elements cannot be split up into simpler substances
- elements all have one-word names (and a symbol). They are found in the periodic table.

Other metals such as iron and aluminium are also elements. Carbon, oxygen and bromine are examples of non-metal elements.

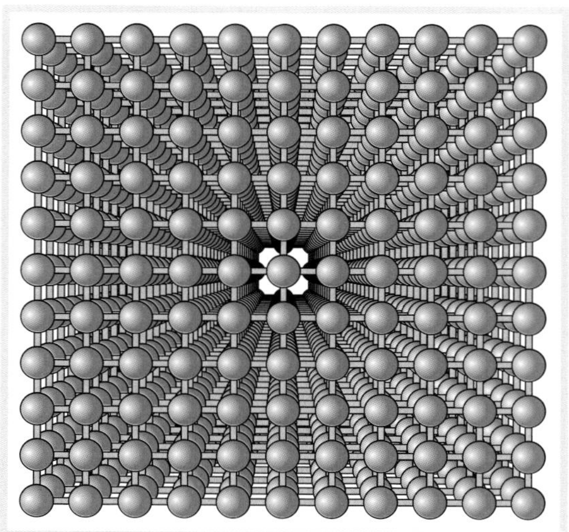

The structure of a solid element. It contains only one type of atom.

Compounds

Carbon dioxide is a **compound** of the elements carbon and oxygen. The carbon and oxygen are bonded together and *a lot* of energy is needed if you want to separate them. You can identify compounds by remembering that:

- compounds contain two or more different types of atom bonded together
- compounds can only be split up into simpler substances by chemical reactions.

The name or the formula of a compound shows you what elements it contains.

Activity A

Look at this list of names of substances and say whether you think they are elements or compounds: **(a)** nitrogen, **(b)** magnesium oxide, **(c)** hydrochloric acid, **(d)** gold.

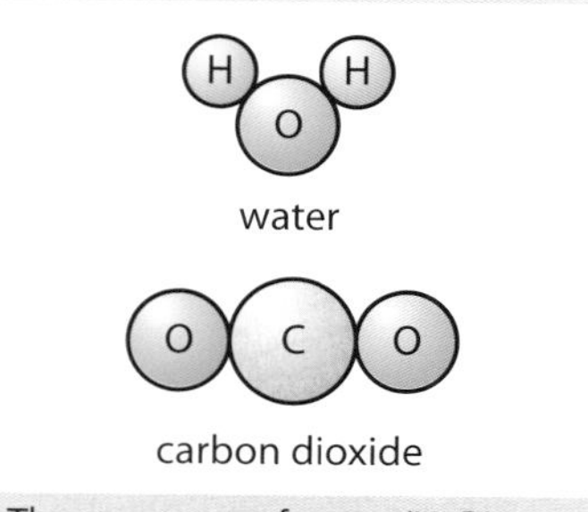

The structures of water (H_2O) and carbon dioxide (CO_2).

When elements join together to make compounds, you can combine their symbols to make a chemical formula. The chemical formula shows you the number of atoms of each element present in the compound. Some examples of compounds are sodium chloride (NaCl), ammonia (NH_3) and sodium hydroxide (NaOH).

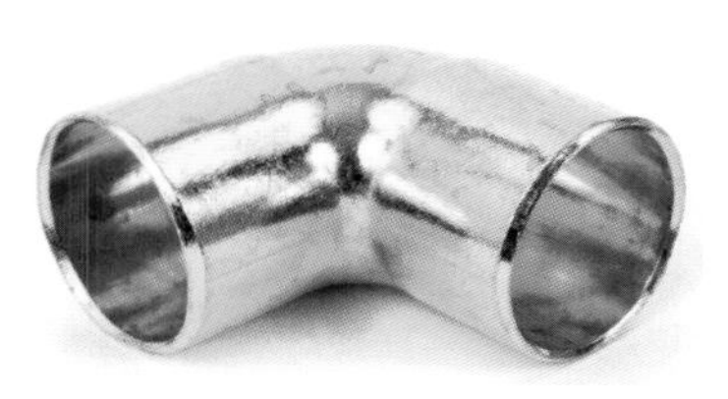

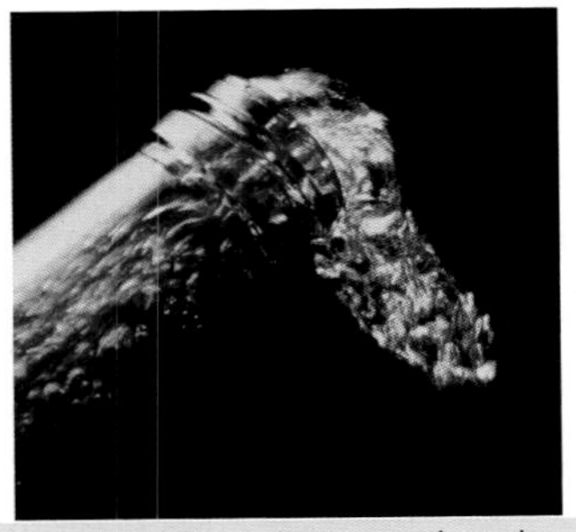

Copper, carbon dioxide and crude oil are all familiar substances, but three different words are used to describe them.

Molecules

Nitrogen, oxygen and chlorine are elements. The formulae of these elements are written as N_2, O_2 and Cl_2. This shows you that the atoms in these elements always go around in pairs. If two or more atoms are bonded together this is called a **molecule**, so these are molecular elements. Compounds such as carbon dioxide are also molecules.

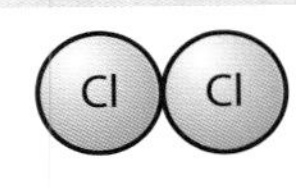

The structure of chlorine (Cl_2).

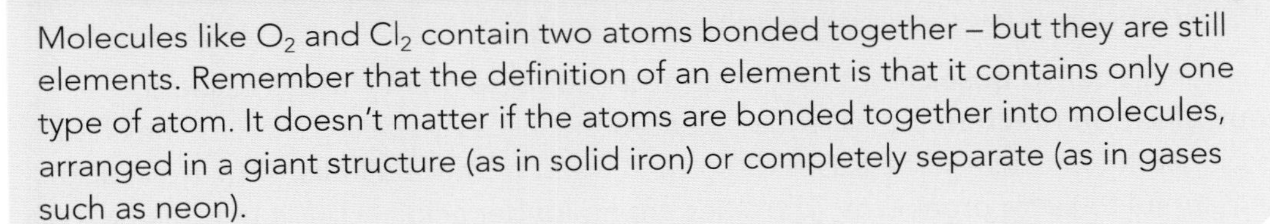

Common misconception

Molecules like O_2 and Cl_2 contain two atoms bonded together – but they are still elements. Remember that the definition of an element is that it contains only one type of atom. It doesn't matter if the atoms are bonded together into molecules, arranged in a giant structure (as in solid iron) or completely separate (as in gases such as neon).

Mixtures

Crude oil contains a **mixture** of different substances, with names such as octane, cyclohexane and naphthalene.

You can identify mixtures by remembering that:

- mixtures contain several different substances that are not bonded together
- mixtures can quite easily be split up into simpler substances using physical processes like **filtration**, **evaporation** and **distillation**, because each substance in the mixture has different physical properties.

The names of mixtures often sound quite familiar, but do not give very much information about what they contain. Examples of mixtures are brine (salt and water) and air (mainly nitrogen and oxygen).

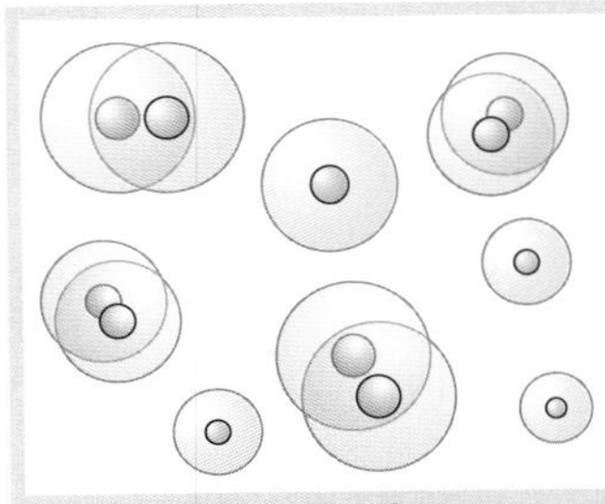

There are two different substances in this mixture, but they are not bonded together.

Just checking

1 'Air contains mostly nitrogen and oxygen, along with small amounts of other substances such as water vapour and carbon dioxide.'

(a) Five substances are mentioned in this sentence. Write down their names and decide whether each is an element, compound or mixture.

(b) The formula of water is H_2O. What can you tell about a molecule of water from this formula?

Lesson outcomes

You should:

- be able to use the periodic table to recognise elements and formulae of simple compounds
- know the definitions of elements, compounds, mixtures and molecules (molecular elements).

1.15 Neutralisation reactions

Key terms

Acid – A solution with a pH of less than 7.

Base – A substance that can neutralise an acid. Bases are often metal oxides or metal hydroxides.

Alkali – A base that is soluble in water. Alkalis have a pH of greater than 7.

Neutralisation reaction – A reaction in which an acid reacts with a base to form water and a neutral salt.

An acid problem

In the late 1960s, a scientist in Sweden noticed that many of the country's lakes had changed dramatically. They had become acidic, which meant that many of the plants and animals that lived in the lakes could no longer survive.

The problem had been caused by **acid rain**. The rain that fell on Sweden was polluted by acidic gases such as oxides of sulfur and nitrogen. In water, these gases form acids such as sulfuric acid and nitric acid. The rainwater drained into the lakes which meant that they contained a lot of sulfuric acid.

Acid pollution causes major environmental problems.

Activity A

Do some research on the Internet and in books to find out more about acid rain.

1 What other countries besides Sweden have had problems with acid lakes?

2 What other effects does acid rain have, besides polluting lakes?

Substances called **bases** react with acids and neutralise them – this is called a **neutralisation reaction**. So, scientists treated the lakes by adding a base. They had to be careful because some bases dissolve in water to make **alkalis**. Alkalis can also be harmful to living organisms. The lakes are no longer acidic – but it may take a long time before the plants and animals return.

Here is a list of common bases.

Name	Formula	Does it dissolve in water to make an alkali?
Calcium hydroxide	$Ca(OH)_2$	Slightly
Copper oxide	CuO	No
Zinc oxide	ZnO	No
Sodium hydroxide	$NaOH$	Yes

Assessment tip

You need to know the formulae of copper oxide, zinc oxide and sodium hydroxide.

Acids and pH

You can test if a **solution** is an acid or an alkali by finding its pH. The lower the pH, the more acidic the solution is.

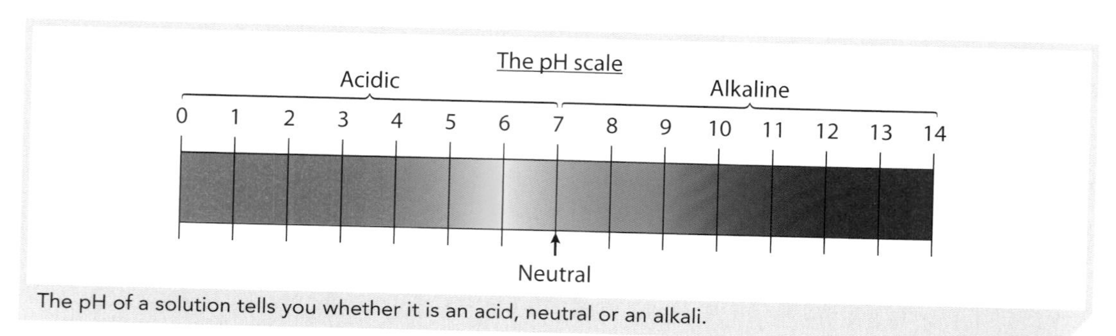

The pH of a solution tells you whether it is an acid, neutral or an alkali.

Indicators

Indicators are chemical substances which change colour depending on the pH of the solution. Litmus is a simple indicator. It turns red in acid and blue in alkali.

Universal indicator is actually a mixture of several indicators. The exact shade of colour can be used to find the pH of the solution, by comparing it with a colour chart.

Common misconception

Litmus has only two possible colours – red and blue. It does not turn a different colour in a neutral solution, so it is not very useful for telling if a solution has a neutral pH.

Case study

Sven is an ecologist in Norway, where there are also problems with acid lakes.

'Our lakes are very beautiful and the public are very angry that they have become polluted by acid rain. About 30 years ago, the pH of many lakes was about 4.0 or even less.

Now, most of the lakes I monitor have a pH of about 5.5. We think that this is because much less sulfur dioxide is being emitted by industries in the rest of Europe. Most of Norway's winds come from the west, blowing in pollution from other countries.

If a lake remains very acidic, I may recommend liming it. This means adding calcium oxide or calcium carbonate. It can help clean up the lake, but can't bring the fish back.'

This is how the lake would be limed.

1. Which pH is more acidic: 4.0 or 5.5?
2. Explain why adding calcium oxide will help change the pH back to 'normal'.
3. Look on an atlas and suggest the names of some countries which may be producing the sulfur dioxide pollution that has affected Norway's lakes.

Just checking

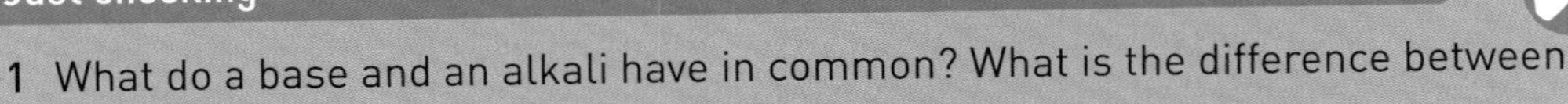

1. What do a base and an alkali have in common? What is the difference between them?
2. Give the formula of **(a)** sodium hydroxide, **(b)** copper oxide.

Dipping a piece of pH paper into a solution gives you information about the pH of the solution.

Activity B

Predict the pH of these substances: **(a)** water, **(b)** a solution of sodium hydroxide, **(c)** sulfur dioxide dissolved in water.

Link

You will learn about neutralising soil in lesson 1.19.

Lesson outcomes

You should:

- know the definitions of acid, base and alkali
- understand how a neutralisation reaction can be used to reduce the acidity of lakes caused by acid rain
- know about testing for pH using universal indicator and litmus.

1.16 Acids and salts

Different acids

Did you know?

Your stomach contains hydrochloric acid and so it has a pH of about 2.

Link

You learned about acid rain in lesson 1.15.

Many natural substances are acids, including ethanoic acid, which is found in vinegar, citric acid, which is found in fruit, and formic acid, which is present in the stings and bites of some insects, such as ants and bees. Gases can also be acidic. Gases such as sulfur dioxide, sulfur trioxide and oxides of nitrogen can cause acid rain.

Sulfuric acid and hydrochloric acid are often found in the clouds of steam around volcanoes.

Case study

Elizabeth is a chemical engineer who works for a national energy generating company.

'We produce energy by burning coal. The heat is used to make steam, which turns the **turbines**. The turbines turn the electrical generators, which convert rotational kinetic energy into electrical energy.

It's important to try and make our process as clean and non-polluting as possible. One thing we have done something about is the release of acid gases like sulfur oxides and nitrogen oxides. The coal we burn contains some sulfur impurities, so when it burns it forms sulfur oxides. We've installed special 'scrubbers' in the chimneys to react with the sulfur oxides and turn them into a harmless salt – calcium sulfate. We can even sell this to other industries.

Also, by burning the coal at a lower temperature we can stop the nitrogen and oxygen in the air from reacting together to form nitrogen oxides. We've reduced our emissions of acid gases by 50% in the last 10 years. This is good for the environment and saves us money by making sure we don't miss our emissions targets.'

1. Sulfur oxides and nitrogen oxides can cause acid rain. Explain how they do this.
2. Elizabeth writes the formula of the sulfur oxides SO_x. Why do you think this is?
3. Suggest what substance could be reacted with the sulfur oxide gases in the scrubber to make calcium sulfate. Remember that calcium sulfate is a salt.
4. Find out one other environmental problem caused by emissions of nitrogen oxides.

Different salts

When an acid and a base react in a neutralisation reaction, the products formed are a salt and water.

The neutralisation products have different properties from the acid and the base that formed them. Water is a neutral substance and so are most salts. In the reaction, the base is added to the acid and the pH gradually becomes 7 (neutral).

The salt that is formed in a neutralisation reaction depends on the acid and the base that react together. The table gives some examples.

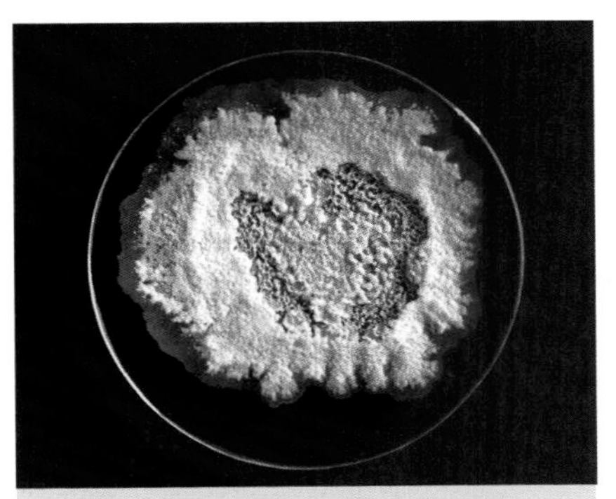

The end product of a neutralisation reaction is a salt. If you evaporate the water away you will leave behind the salt. All salts are solid crystals at room temperature.

Assessment tip

You will need to know the formulae of sulfuric acid, hydrochloric acid and nitric acid and be able to use them to write balanced equations for the reactions you have studied.

Predicting the products

The name of the salt shows you what acid has been used to form it.

- Sulfates are formed from sulfuric acid.
- Chlorides are formed from hydrochloric acid.
- Nitrates are formed from nitric acid.

Acid	Formula	Base	Salt formed
Sulfuric acid	H_2SO_4	Copper oxide	Copper sulfate
Hydrochloric acid	HCl	Sodium hydroxide	Sodium chloride
Nitric acid	HNO_3	Zinc oxide	Zinc nitrate

Activity A

Predict the salts formed by these reactions between an acid and a base:

- hydrochloric acid and sodium hydroxide
- nitric acid and copper oxide
- sulfuric acid and zinc oxide.

The name of the salt also shows you what metal was present in the base. This also means that you can predict what salt will be formed if you know the names of the acid and base that react together.

The salt copper chloride is formed when copper oxide and hydrochloric acid react.

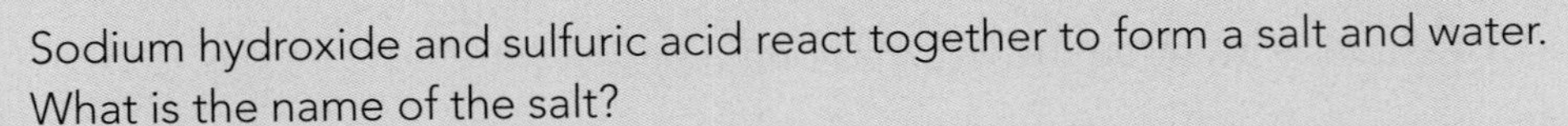

Worked example

Sodium hydroxide and sulfuric acid react together to form a salt and water. What is the name of the salt?

Step 1 The metal in the base, sodium hydroxide, is sodium.

Step 2 Sulfuric acid forms sulfate salts.

Step 3 So the full name of the salt formed will be sodium sulfate.

Lesson outcomes

You should:

- know about neutralisation reactions between hydrochloric acid, nitric acid or sulfuric acid, and a metal oxide or a metal hydroxide
- understand that the products of a reaction will have different properties from the reactants.

1.17 Equations for neutralisation reactions

Key term

Word equation – A way of describing what happens in a chemical reaction. Word equations show you the names of the substances that react together and the new products formed.

Word equations

To predict what products will be formed in a reaction, we need to know how to write **word equations**.

We can describe what happens in a neutralisation reaction with a general word equation:

acid + base → a salt + water

Activity A

Find the table in lesson 1.16 that shows the salts formed from three different combinations of acid and base. Write word equations for the three neutralisation reactions. Remember that water is also formed in each of the reactions.

Assessment tip

You may be asked to balance an equation, or to write down the balanced equation for a specific problem. But the equations will all be connected with neutralisation reactions so with practice you will soon get used to this kind of problem.

Formulae and balanced chemical equations

Chemists often prefer to write balanced chemical equations for reactions. These give more information about the reactions:

- the formulae of the substances involved in the reaction
- the numbers of particles of each substance involved in the reaction.

When we write balanced chemical equations, we need to make sure that all the atoms appear on both sides of the equation. The balanced chemical equation for the reaction of sodium hydroxide with hydrochloric acid is:

sodium hydroxide	+	hydrochloric acid	→	sodium chloride	+	water
NaOH	+	HCl	→	NaCl	+	H_2O

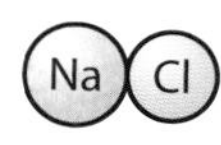

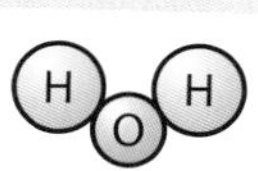

Drawing diagrams of the formulae in an equation helps you to see whether the chemical equation is balanced.

Remember

Don't forget that the small '2' in the formula H_2O indicates that there are two hydrogen atoms in the water molecule. In any chemical formula the small numbers refer to the atom to the left. So the molecule H_2SO_4 has two hydrogen atoms and four oxygen atoms.

This equation shows that the chemical formula for sodium hydroxide is NaOH. We know that the equation above is balanced because:

- there is one sodium atom on the left hand side, and one sodium atom on the right hand side of the arrow
- there is one chlorine atom on the left, and one on the right
- there is one oxygen on the left, and one on the right
- there are two hydrogen atoms on the left, and two on the right (remember, the little '2' after the 'H' in H_2O means that there are two hydrogen atoms).

Another example of a neutralisation reaction is when sulfuric acid and sodium hydroxide react together. The balanced chemical equation for this is:

sulfuric acid	+	sodium hydoxide	→	sodium sulfate	+	water
H_2SO_4	+	2NaOH	→	Na_2SO_4	+	$2H_2O$

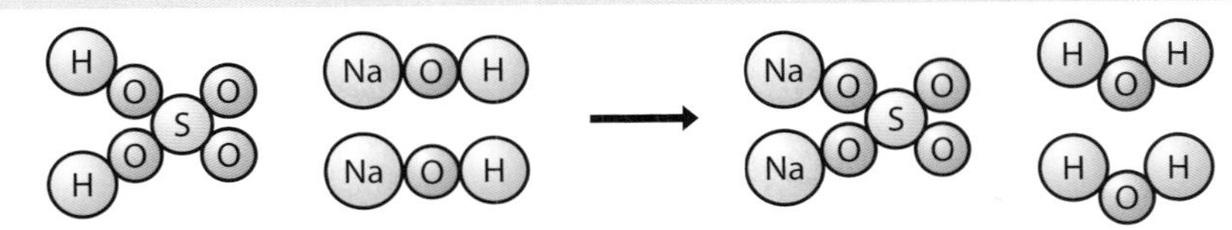

A diagram of the reaction between sulfuric acid and sodium hydroxide.

In this reaction, there are two sodium atoms on the right hand side of the equation which are in sodium sulfate (Na_2SO_4). But sodium hydroxide (NaOH) on the left hand side only contains one sodium atom. Therefore, to balance the equation, there must be two molecules of sodium hydroxide, which is written as '2NaOH'. It means that there are two of every atom in the molecule, so there are two sodium atoms, two oxygen atoms and two hydrogen atoms.

The same idea explains why there is a 2 in front of the formula for water – it makes sure that the numbers of hydrogen and oxygen atoms are the same on the two sides of the equation.

Link

You will learn more about balanced chemical equations in lesson 2.8.

Worked example

Copper oxide reacts with hydrochloric acid to form copper chloride and water. Write the word equation and the balanced chemical equation for this reaction.

The word equation is:

copper oxide + hydrochloric acid → copper chloride + water

Follow these steps to write a balanced chemical equation.

Step 1 Look up (or remember) the chemical formula of each of the substances involved in the reaction. The formula of copper chloride is $CuCl_2$, the formula of water is H_2O and the other formulae can be found in lessons 1.15 and 1.16.

Step 2 Replace the names in the word equation by the formulae of the substances:

$$CuO + HCl \rightarrow CuCl_2 + H_2O$$

Step 3 Check to see whether any of the elements need to be balanced. In this case, there is 1 Cl and 1 H atom on the left hand side and 2 Cl and 2 H on the right-hand side. This means you need to write a 2 in front of the HCl formula:

$$CuO + 2HCl \rightarrow CuCl_2 + H_2O$$

Remember

The reaction between sulfuric acid and sodium hydroxide produces two water molecules. But remember that each water molecule has two hydrogen atoms – H_2O – making four hydrogen atoms altogether.

Just checking

1 Sulfuric acid reacts with copper oxide in a neutralisation reaction to form water and one other product.
  (a) What is the name of the other product?
  (b) Write a word equation for this reaction.

2 Copy and complete this equation for a neutralisation reaction by adding the correct formula and balancing number:

$$\_\_\_HCl + \_\_\_\_ \rightarrow ZnCl_2 + H_2O$$

Lesson outcomes

You should know how to write word equations and simple balanced chemical equations for the reactions in this unit, using the correct chemical formulae.

1.18 Acids and metals

Brass (an alloy of copper and zinc) might look like gold but it fails the acid test – it will fizz when acid is dropped onto it.

The acid test

Gold is one of the most valuable metals on Earth. But not all golden metals are gold – you could easily be fooled into thinking that the yellow, shiny alloy called brass was gold. In years gone by, people have lost a lot of money paying high prices for worthless minerals or alloys that looked just like gold.

The answer to this problem was to use the 'acid test'. If a few drops of acid are dripped onto gold, nothing happens. Gold is one of the most unreactive metals of all. But with most other metals, you would see fizzing and bubbling. The metals react with the acid and produce a gas.

Remember

Hydrogen gas has the chemical formula H_2 because it is made up of molecules containing two hydrogen atoms bonded together.

Acids and metals

When metals react with acids, a salt is formed as in neutralisation reactions. But what is different is that the other product is not water, it is hydrogen gas.

So the general equation is:

acid + metal → salt + hydrogen

Some examples:

calcium + sulfuric acid → calcium sulfate + hydrogen

$$Ca + H_2SO_4 \rightarrow CaSO_4 + H_2$$

magnesium + hydrochloric acid → magnesium chloride + hydrogen

$$Mg + 2HCl \rightarrow MgCl_2 + H_2$$

Worked example 1

Write the word equation for the reaction of zinc with hydrochloric acid.

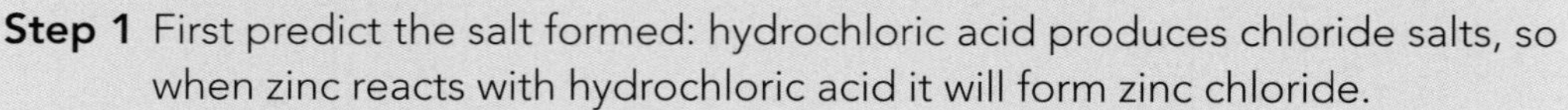

Step 1 First predict the salt formed: hydrochloric acid produces chloride salts, so when zinc reacts with hydrochloric acid it will form zinc chloride.

Step 2 The other product is hydrogen.

Step 3 So the word equation is:

zinc + hydrochloric acid → zinc chloride + hydrogen

Worked example 2

Copy and complete this balanced chemical equation by filling in the missing formulae and balancing the numbers:

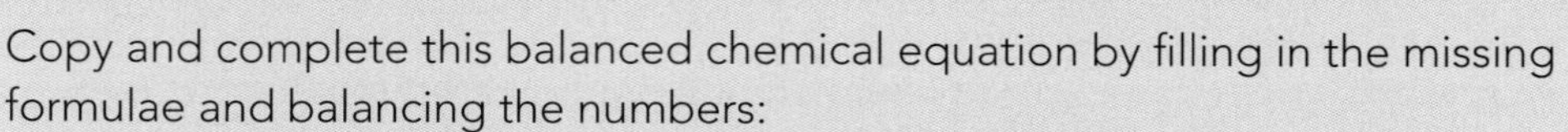

$$Mg + \_\_\_\_\_\_ \rightarrow MgCl_2 + \_\_\_\_$$

Step 1 Metals react with acids to form hydrogen. So the missing product is H_2.

Step 2 The other product is magnesium chloride. So the acid must be hydrochloric acid, HCl.

Step 3 There are two Cl atoms and two H atoms on the right hand side of the equation. So there must be two HCl molecules:

$$Mg + 2HCl \rightarrow MgCl_2 + H_2$$

Safety and hazards

Never add concentrated nitric acid to a metal. A different kind of reaction happens that produces nitrogen dioxide. Nitrogen dioxide is a red-brown gas, which is toxic (poisonous) if you breathe it in.

metal + nitric acid → metal nitrate + water + nitrogen dioxide

For example:

copper + nitric acid → copper nitrate + water + nitrogen dioxide

$Cu + 4HNO_3 \rightarrow Cu(NO_3) + 2H_2O + 2NO_2$

Activity A

1 Copy and complete this word equation:

calcium + sulfuric acid → ____________ + __________

2 Copy and complete this balanced equation by adding the missing formula and balancing number:

____ + ___ HCl → $ZnCl_2 + H_2$

Testing for hydrogen

If you see that a gas is being produced in a reaction you can check that it is hydrogen gas by carrying out a simple test. The test works best if the reaction is happening in a test tube. Place a cork loosely over the top of the test tube for a few seconds to let the gas build up. Then light a wooden splint, remove the cork and hold the flame next to the mouth of the test tube. If the gas in the test tube is hydrogen you will hear a squeaky 'pop'.

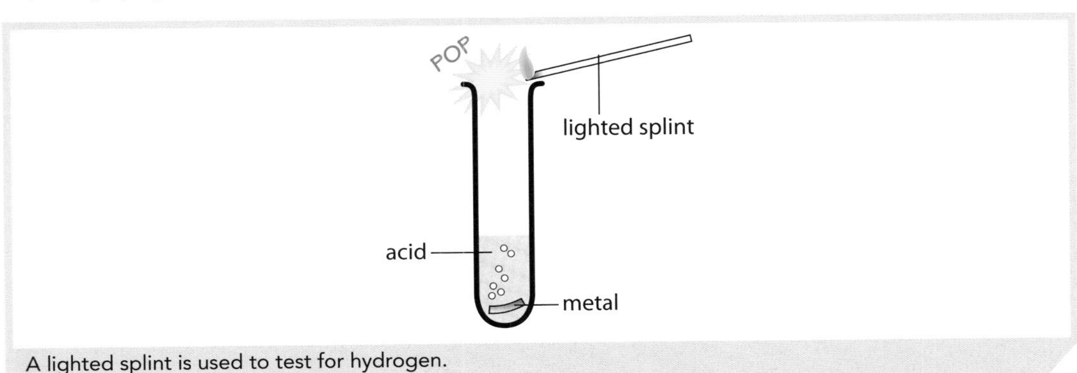

A lighted splint is used to test for hydrogen.

Just checking

1 What products are formed when magnesium reacts with sulfuric acid?

2 What is the formula of hydrogen gas?

Did you know?

You hear a pop in the hydrogen test because a small explosion is happening in the test tube. The hydrogen is reacting very quickly with oxygen from the air to make water vapour.

Lesson outcomes

You should know:

- the reactions of hydrochloric acid and sulfuric acid with metals
- how to write the word and balanced chemical equations for these reactions
- the chemical test for hydrogen.

1.19 Acids and carbonates

Get started

Find out the pH of the soil in the grounds of your school or college. Take about 10 g of soil (about a spoonful) and add it to 50 cm^3 of water in a measuring cylinder. Shake the soil and water together and let the soil settle to the bottom of the measuring cylinder. Then dip a piece of litmus paper into the water at the top of the measuring cylinder. Is the soil acidic or basic?

Link

You have learnt about liming lakes in lesson 1.15.

Neutralising acid soils

Farmers know the importance of checking the pH of their soil. A lot of the crops that they grow, like wheat and maize, grow best at a pH of about 6. But soils can easily become more acidic, perhaps when fertiliser or compost is added, or because of the effects of acid rain.

To try and reduce the acidity of soil, farmers could add bases like calcium oxide. But this is quite a harmful substance to handle – it can cause burns or breathing difficulties, and reacts with water to form an alkali.

A much safer and environmentally friendly way of reducing acidity is to use calcium carbonate. This occurs naturally as limestone and chalk. It is ground up into a fine powder and ploughed into the soil where it slowly reacts with the acids.

Adding calcium carbonate to the soil is called liming.

Acids and carbonates

Carbonates are compounds containing a metal and a carbonate group, CO_3.

Name	Formula
Sodium carbonate	Na_2CO_3
Copper carbonate	$CuCO_3$
Calcium carbonate	$CaCO_3$

When carbonates react with acids, a salt and water are formed, as in neutralisation reactions. But there is another product as well – carbon dioxide gas. So the general equation is:

acid + carbonate → salt + water + carbon dioxide

Assessment tip

You need to know the formulae of these three carbonates. Remember that the carbonate group has the formula CO_3. You can then look up the symbol for the metal in the periodic table – but remember that the formula of sodium carbonate has two sodiums to every carbonate group.

Some examples:

hydrochloric acid + calcium carbonate → calcium chloride + water + carbon dioxide

$2HCl + CaCO_3 \rightarrow CaCl_2 + H_2O + CO_2$

sulfuric acid + sodium carbonate → sodium sulfate + water + carbon dioxide

$H_2SO_4 + Na_2CO_3 \rightarrow Na_2SO_4 + H_2O + CO_2$

nitric acid + copper carbonate → copper nitrate + water + carbon dioxide

$2HNO_3 + CuCO_3 \rightarrow Cu(NO_3)_2 + H_2O + CO_2$

Activity A

Write word equations to describe what happens when:

1. hydrochloric acid is added to sodium carbonate
2. sulfuric acid is added to copper carbonate.

Activity B

Look at these three equations. Are they balanced? If not, copy them out and balance them by writing numbers in the appropriate place.

1. $HCl + CuCO_3 \rightarrow CuCl_2 + H_2O + CO_2$
2. $H_2SO_4 + CaCO_3 \rightarrow CaSO_4 + H_2O + CO_2$
3. $HNO_3 + Na_2CO_3 \rightarrow NaNO_3 + H_2O + CO_2$

Testing for carbon dioxide

When carbonates react with acids, you will see a gas being produced. To check that the gas is carbon dioxide, you can carry out a test by bubbling the gas through limewater (a solution of calcium hydroxide). If the gas is carbon dioxide then you will see the limewater turn cloudy after a few seconds.

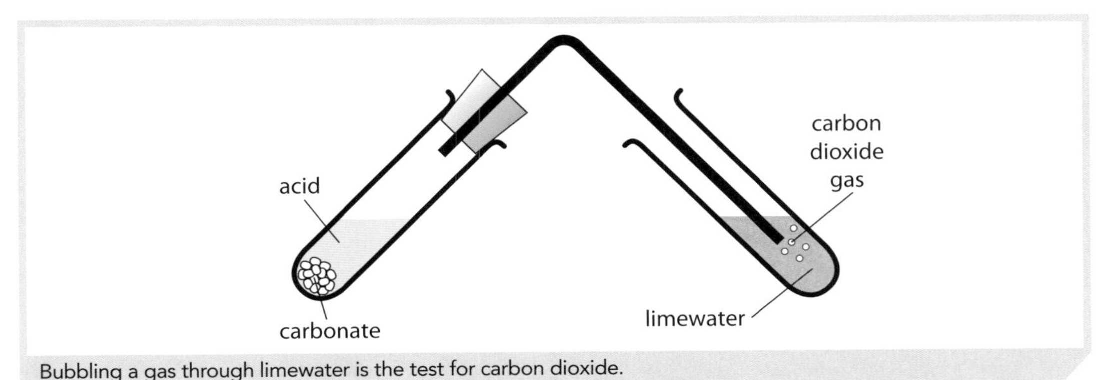

Bubbling a gas through limewater is the test for carbon dioxide.

Lesson outcomes

You should:

- know the reactions of hydrochloric acid, sulfuric acid and nitric acid with sodium carbonate, copper carbonate and calcium carbonate
- know how to write the word and balanced chemical equations for these reactions
- know the chemical test for carbon dioxide
- understand how neutralisation reactions are used to reduce the acidity of soils.

Just checking

1. Ajay tests an unknown substance. He reacts it with hydrochloric acid. A gas is given off which turns limewater cloudy.
   - **(a)** What is the gas?
   - **(b)** Look at the names of these three substances: **(a)** magnesium, **(b)** sodium hydroxide, **(c)** calcium carbonate. Which one is most likely to be the unknown substance?
2. Farmers sometimes spread calcium carbonate onto soils. Explain why they do this.

1.20 Hazards of acids and bases

Your hazardous stomach

Your stomach is naturally acidic – it contains hydrochloric acid and has a pH of around 2. Some people suffer from reflux, when stomach acid escapes into their oesophagus. The stomach acid is harmful – it causes damage to the cells of the lining of the throat. The stomach itself has a special lining that protects it against the hydrochloric acid.

Acids like hydrochloric acid are always labelled in the laboratory to warn you of their hazards.

Hydrochloric acid is harmful – it will cause health problems if it is swallowed or absorbed through the skin.

Settling the stomach

People who suffer from reflux, or stomachs that make too much acid, will often take indigestion remedies to relieve the problem. Indigestion remedies are powders, or mixtures of powder and water, that will neutralise this extra acid. Choosing the right substance to neutralise the acid is very important. The indigestion remedy must have as little hazard as possible.

Indigestion

Indigestion remedies contain bases or carbonates, either by themselves or combined together:

- calcium carbonate, $CaCO_3$
- magnesium hydroxide, $Mg(OH)_2$
- aluminium hydroxide, $Al(OH)_3$.

If you look at containers of these substances, you will see that they do not carry a hazard symbol. But does that mean they are safe?

Swallowing large quantities of these substances can cause a range of problems.

Substance	Possible risks
Calcium carbonate	Kidney stones Bone pain Build up of carbon dioxide gas
Magnesium hydroxide	Laxative effects
Aluminium hydroxide	Constipation Possibly toxic to brain cells

It is important to take note of the safety information on the packaging of any medicine.

The packaging of indigestion remedies will show the recommended daily safe dose. It is important not to take more of the remedy than this limit. Even if the patient takes less than the daily safe dose, taking the remedy over a long period of time could still cause problems.

Activity A

Sodium hydroxide can neutralise acids. Give one reason why it is not used as an ingredient in indigestion remedies.

Other hazards

Many of the substances you have studied in this unit have particular hazards. Any bottle or container of these substances must be labelled with a hazard symbol to indicate in what way it is hazardous.

Sodium hydroxide

Solid sodium hydroxide reacts very rapidly with acids. But it is also extremely corrosive – even when dissolved in water as an alkali. It will rapidly cause burns and blisters to skin or other tissue.

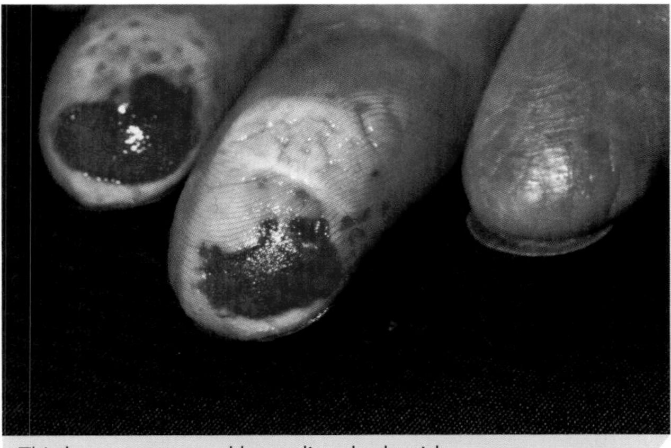
This burn was caused by sodium hydroxide.

Copper oxide

Copper oxide is an irritant – this means that it will cause reddening or slight blistering of the skin.

If copper oxide were swallowed it would also be toxic. This often isn't mentioned on hazard symbols, because normally there is very little risk of this happening.

However, if large amounts of copper oxide are dumped in the ground or washed away into drains, there is a risk that this could damage the environment by poisoning plants or animals.

Magnesium

If you used magnesium metal in practical work in this lesson, you probably used it as a thin ribbon of metal. In this form it is very flammable and so is the hydrogen which is formed when magnesium reacts with acids. It will start burning if it is accidentally ignited by a flame.

Sodium hydroxide is a corrosive substance.

Copper oxide is harmful and hazardous to the environment.

Magnesium metal is flammable.

Safety and hazards

If a substance is hazardous, it is essential that you find out more about the **hazards** and how to reduce the **risk** they pose. Anyone using hazardous substances needs to carry out a risk assessment to show that they have taken all possible steps to make sure that they are using the substances safely.

Did you know?

When substances are dissolved in water to make solutions, the hazards can change depending on the concentration of the solution. If the solution has a low concentration, it may be much less hazardous than a solution with a high concentration.

Activity B

Use the Internet to find out the hazards (if any) of these substances which you have met in this unit: (**a**) calcium oxide, (**b**) copper carbonate, (**c**) zinc oxide, (**d**) carbon dioxide.

Lesson outcomes

You should:

- know the hazard symbols for the chemicals used in this unit
- understand the application of neutralisation reactions in indigestion remedies.

1.21 Energy and its use

Key term

Joule – The joule (symbol J) is the scientific unit used to measure energy.

Energy is incredibly important in the modern world. Huge power stations produce enough electrical energy to power whole cities. Engines provide **mechanical energy** for cars, trucks, trains, planes and all kinds of machinery. Boilers and other kinds of heater produce **thermal energy** to heat homes and workplaces. And the Sun is the biggest energy source of all, pouring out vast amounts of light energy and other **radiation**.

Energy is the ability of a system to do work. You could say that it is the ability to make things happen. Whenever something happens, anywhere in the Universe, energy is involved in making it happen.

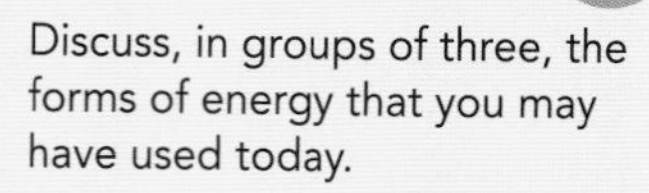

Discussion point

Discuss, in groups of three, the forms of energy that you may have used today.

Electrical energy provides power for lights, which give out light energy.

Energy comes in many forms – light, thermal energy, sound, movement, electricity and **nuclear energy**. Mechanical energy is the energy an object has either because it is moving or because of its position. The energy of a moving object is **kinetic energy**. But an object can also have **potential energy**, which is energy that is available but is not being used. A book on a shelf, for example, has potential energy that could become kinetic energy if it fell.

A military plane approaching the speed of sound. The aircraft has kinetic energy. As it breaks the sound barrier it produces a sonic boom, which is sound energy.

A type of image called a thermogram shows the thermal energy radiating from a house. The brighter areas are giving out more energy.

Stored energy

Car engines, heaters, electric generators and living things all need fuel to make them work. A fuel is a source of stored energy. Energy can be stored in various ways.

- Food is a store of **chemical energy**. Drinks can be chemical energy stores too. In the body, this chemical energy is used to move, grow, build and repair tissue, and to keep warm.
- Petrol, diesel and other fuels used in vehicles and in heaters are also stores of chemical energy. In a car engine, the chemical energy is turned into kinetic energy.
- The water in a reservoir behind a dam will have stored **gravitational potential energy**. When water is released and falls from the top of the dam, this is transformed into kinetic energy.
- If you slowly pull back a catapult or a bowstring, you store **elastic potential energy**. Let go, and the potential energy becomes kinetic energy. Rubber, springs or any elastic material store potential energy when they are stretched or compressed.
- Thermal energy comes from the movement of the atoms that everything is made of, so even cold objects have some thermal energy. If an object is warmer than its surroundings, thermal energy will flow from the warmer object to the cooler surroundings.
- The nucleus of an atom can be a store of energy. When the nucleus of a large, unstable atom such as uranium is split, the stored energy is released. This can happen gradually, as in a nuclear reactor, or in a sudden explosion, as in a nuclear weapon.

The fuel rods in a nuclear power station are a store of nuclear energy. In the reactor this is converted to thermal energy, which heats water to steam. The steam is used to turn turbines.

Water held back by a dam is a store of gravitational potential energy. If some water is allowed to flow along pipes to the base of the dam, the potential energy becomes kinetic energy that can be used to turn turbines.

Just checking

1. What are the forms of energy involved when a car horn is operated?
2. Where can nuclear energy be used, other than in a nuclear power station for generating electricity?
3. What form of energy does a person sitting on a hill have?
4. What form of energy is released by splitting large unstable atoms?
5. What form of energy is stored in the muscles of humans?

Did you know?

A typical apple stores 150 000 **joules** of energy.

Link

You will learn more about how nuclear energy can be released in lesson 3.4.

Activity A

1. Using the Internet and books, investigate how much energy is used by these organs when your body is at rest:
   (a) the brain
   (b) the liver
   (c) the heart.
2. Which of these organs uses the most energy?
3. Although the joule (J) is the unit of energy, calories are often used to measure the energy stored in food. For example, a large carrot is estimated to store about 30 kilocalories (30 000 calories) of chemical energy. Investigate how to convert this into joules. What is 30 kilocalories in kilojoules (kJ)?

Lesson outcomes

You should be able to describe the various forms of energy, their uses and how energy can be stored.

1.22 Energy transformations and transfers

Key terms

Transformation – When energy changes from one form to another.

Electric charge – Charged particles that transfer electrical energy in electrical devices.

Energy transformation diagrams

Energy is never lost, but can be transformed from one form to another. This **transformation** can be represented by rectangular boxes in a diagram. The images show how energy is transformed in four different applications: a reading lamp, boiling a kettle to make tea, a book falling off a shelf and a battery-operated cooling fan.

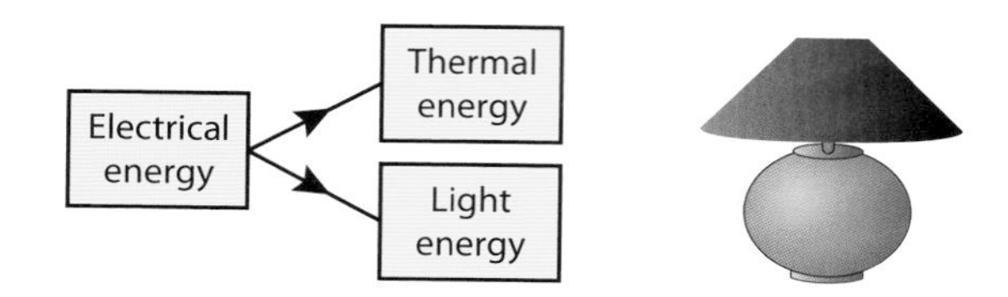

When a light bulb is turned on, electrical energy is transformed into light energy. This 'useful energy' is the energy we want out of the light bulb. However, you may have noticed that light bulbs get hot. This means that some of the electrical energy has also been transformed into thermal energy.

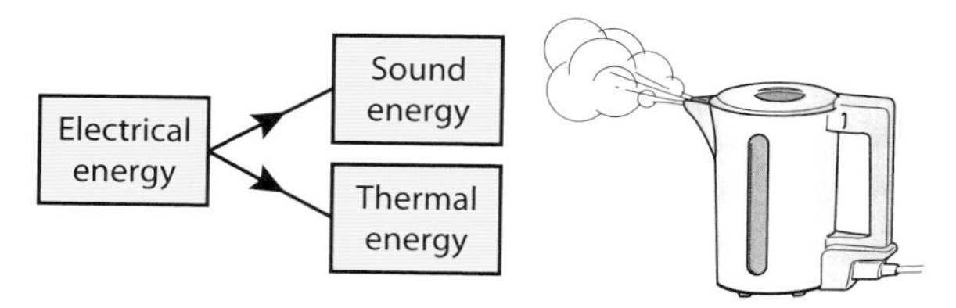

In an electric kettle, electrical energy is transformed into thermal energy in the heating element. The thermal energy transfers to the water. When the water boils, a little of the energy is transformed into sound.

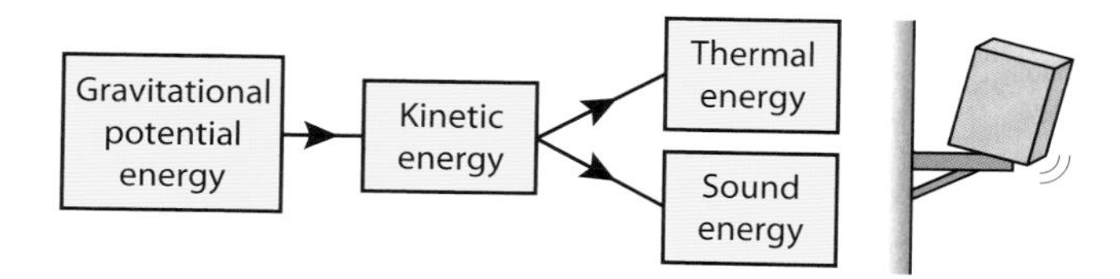

If a book or other object is on a shelf, it has gravitational potential energy. If it falls off the shelf, the gravitational potential energy is transformed into kinetic energy. When the object hits the ground with a crash, some of the kinetic energy transforms into sound energy. The rest becomes thermal energy.

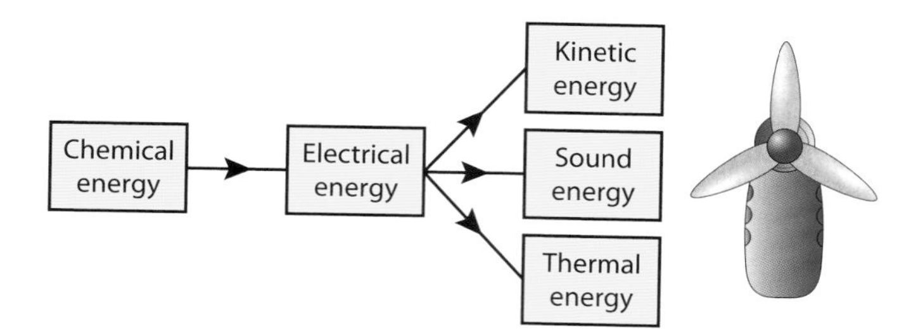

In a battery-operated cooling fan, the battery transforms chemical energy into electrical energy. The fan's motor turns electrical energy into kinetic energy (moving the fan and the air), thermal energy and sound energy.

Activity A

With a partner, discuss the energy transformations in the following examples.

1 Throwing a cricket ball, which hits the ground, rolls and stops.
2 A bus as it travels along the road.
3 Carrying a parcel to a flat on the fourth floor.
4 Blowing a trumpet.

For each example draw an energy transformation diagram and identify what is useful energy and what is wasted.

Energy transfers

Mechanical transfer

When you start a car, the engine begins transforming chemical energy in the fuel into mechanical energy – the spinning of the car's crankshaft. But when you put the car in gear, there is an **energy *transfer***. The mechanical energy of the crankshaft is transferred to the wheels. The wheels start to turn and the car pulls away. Mechanical energy is transferred when a force causes an object to move over a distance. This transfer of energy is also called work done.

Link

You will learn more about how electrical devices work in lesson 3.7.

How thermal energy is transferred is described in lesson 1.23.

Dragsters are cars designed for a quick getaway. When they start, a lot of energy is suddenly transferred to the wheels.

Electrical transfer

Everyday electrical devices, such as DVD players or TVs, have electrical circuits that allow electrical energy to be transferred. It is the **electric charge** in the circuit that transfers energy from one place to another within the circuit.

Activity B

In groups of three, make a list of how you may have transferred energy before coming to school or college today.

Lesson outcomes

You should understand the different types of energy transformation and how energy is transferred mechanically and electrically.

1.23 Thermal energy transfer

Key terms

Conductor – Material that can conduct heat easily, such as aluminium.

Insulator – Material that is a poor conductor of heat, such as cotton wool or air.

Thermal energy is being transferred all around us. We transfer thermal energy when we cook, or when we switch on a car engine. Engineers who design or install hot-water systems and central heating need to understand thermal energy transfers. An energy transfer is the reason why we feel warmer in the day than at night.

Thermal energy can be transferred in three ways: **conduction**, **convection** and radiation.

Conduction

Conduction happens best in solid materials. In conduction, thermal energy is transferred because there is a temperature difference between one end of the material and the other. The atoms in solid substances vibrate. If one end of a solid is hotter than the other, the atoms in the hotter area have more energy and vibrate more. They jostle against the atoms next to them, and pass on some of their extra energy. Gradually the energy is transferred from atom to atom along the solid.

- If you heat one end of a metal rod, the other end gets hot quite quickly. Thermal energy transfers easily through metals: they are good **conductors** of heat.
- If you heat one end of a glass rod, it takes a long time for the other end to get hot. Energy moves much less easily through materials such as plastic, wood and glass. They are **insulators**. The handles of cooking pans are often made of plastic because it is a good insulator.

Gases, such as air, are very poor conductors. There are gaps between the atoms, so vibrations cannot be passed on. This is why duvets keep us warm. They are full of tiny pockets of trapped air, which stop heat being transferred. A mammal's fur or a bird's feathers work in a similar way. They trap an insulating layer of air close to the body.

Did you know?

Understanding thermal energy transfers allows architects to calculate the heat loss from houses. They can then design heat-efficient homes. For example, heat loss can be reduced by having loft insulation in roofs and by filling wall cavities with insulating material.

Convection

Thermal energy transfer can also occur by convection currents in **fluids** – that is liquids and gases.

- Convection helps a radiator or heater to warm a whole room. Convection currents carry warm air from above the heater to other parts of the room.
- In old domestic hot-water systems, convection currents take energy from the boiler up to a storage tank. This storage tank is then connected to the hot taps.

Convection in fluids can be observed in the laboratory. Put some crystals of potassium permanganate at the bottom of a beaker of water, then heat the beaker with a Bunsen burner.

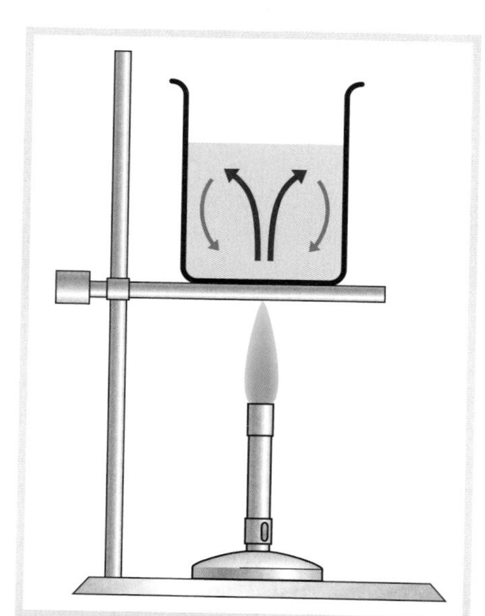

Hot liquids expand, become less dense and rise, then fall as they cool and get more dense. This is called a convection current.

The movement of the convection currents that form can be seen as the water starts to be coloured purple by the crystals. When part of the water is heated, the water molecules move faster and bump into each other more often. The water molecules move further apart and the water expands. This makes the heated water less dense, and it rises. When the water cools, the molecules slow down. The water becomes more dense and it sinks. These movements produce a convection current.

Activity A

Use the ideas about conduction and convection to explain how soup heats up in a pan.

Radiation

Energy can also be transferred by radiation. This is the transfer of energy via electromagnetic waves. Energy from the Sun is transferred by radiation.

When radiation strikes a material, some energy is absorbed by the atoms. The energy makes the atoms vibrate more (in solids) or move faster (in liquids or gases). The material gets hotter. This type of radiation is called **infrared**.

The Hubble Space Telescope is covered in highly reflecting foil to reduce temperature differences as the telescope moves in and out of the Sun's rays.

Energy in the form of radiation can travel through a vacuum and doesn't need atoms or molecules for the transfer to happen. This is different from conduction and convection, where the energy transfer happens because of atoms vibrating or moving.

Link

You can learn more about electromagnetic radiation in lesson 1.28.

Dull black surfaces are good absorbers of radiation. They are also good radiators, so they emit a lot of radiation. Cooling fins on the back of fridges and freezers are dull black so that they radiate away more energy.

Bright shiny surfaces are poor absorbers of radiation as they reflect radiation away.

- Firefighters often wear bright, shiny suits that don't absorb too much infrared radiation. The suits stop them getting burned.
- Tennis players tend to wear bright white shirts to reduce the absorption of energy from the Sun.
- Houses in hot countries are often painted white for the same reason.

Lesson outcome

You should be able to explain how thermal energy is transferred, by conduction, convection and radiation.

Key terms

Sankey diagrams – Block diagrams that show how energy is transformed, the width of the arrow representing the amount of energy.

Efficiency – The measure of how much useful energy is output from a system compared to how much energy is put in. It is normally a percentage.

Conservation of energy – This is the principle that energy cannot be created or destroyed. It can only transform from one form to another.

Energy can never be destroyed – instead it is transformed into different forms. This is the principle of the **conservation of energy**. You can show energy transformation in the form of block diagrams. A **Sankey diagram** is one type of block diagram. It shows *how much* energy is transformed into different forms. The width of each arrow represents the relative amount of energy being used.

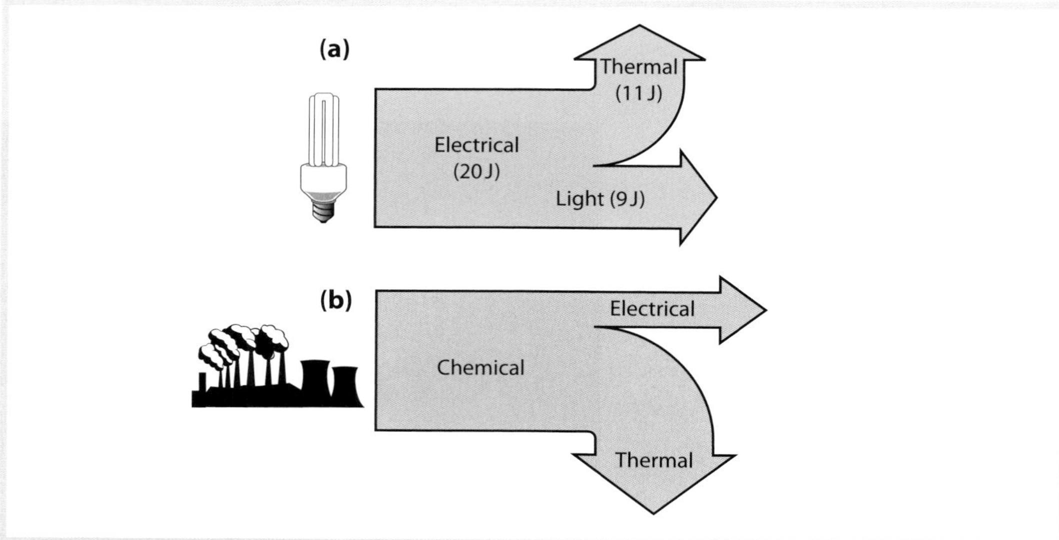

Energy (Sankey) diagrams. **(a)** High-efficiency light bulb: electrical energy transforming to light and thermal energy. **(b)** Power station generating electricity: chemical energy transforming to electrical energy and thermal energy.

The amount of useful (output) energy you get from something like a light bulb or a power station, compared to how much energy you put in (input energy) is the **efficiency** of the system. The efficiency can be calculated using the following equation.

$$\text{efficiency} = \frac{\text{useful energy}}{\text{total energy supplied}} \times 100\%$$

You can calculate efficiency from a Sankey diagram.

Worked example 1

Study the Sankey diagram of the light bulb, and calculate the efficiency.

Step 1 The input energy (electrical) is 20 J and the useful energy (output) is 9 J of light.

Step 2 The efficiency is given by (useful energy ÷ total energy supplied) × 100%:

$$(9 \div 20\,\text{J}) \times 100\% = 45\%.$$

Power

Power is the measure of the amount of energy transferred per unit time, It is measured in watts (symbol W). You can calculate power from the following equation:

$$\text{power (watts)} = \frac{\text{energy (joules)}}{\text{time (seconds)}}$$

Sometimes power is given in kilowatts (for example on the label of electrical kettles). One kilowatt (1 kW) = 1000 W (10^3 W). Power may also be given in megawatts (for example in power stations). One megawatt (1 MW) = one million watts (1 000 000 W or 10^6 W).

Worked example 2

A typical electric kettle is rated at 3 kW. How many joules of energy are transferred in 5 seconds?

Step 1 Write down the equation: power = $\frac{\text{energy}}{\text{time}}$

Step 2 Make energy the subject of the equation: energy = power × time

Step 3 Substitute the numbers into the equation, remembering that 3 kW is 3000 W:

energy = 3000 W × 5 s = 15 000 J (15 kJ)

The cost of electricity

An electricity meter is used to measure how much energy you use. If you look closely you will see that the units of measurement are kilowatt-hours (kWh).

If the power is measured in kilowatts (kW) and time in hours (h), then the energy transferred will be in kWh. You can see this from the equation:

energy transferred (kWh) = power (kW) × time (h)

Household electricity meter. The reading shows that 39 513 kWh of energy has been used.

Worked example 3

A 2 kW electric heater, is used for 4 h. How much does it cost, at 12 p per kWh?

Step 1 Write down the equation: energy transferred (kWh) = power (kW) × time (h)

Step 2 Substitute the numbers into the equation: energy = 2 kW × 4 h = 8 kWh

Step 3 Cost at 12 p per kWh is = 8 kWh × 12 h = 96 p.

Activity A

Investigate how much energy you use at home by keeping an energy diary.

1. Read your electricity meter first thing in the morning and last thing at night.
2. Work out how much energy you have used in kWh.
3. Find out how much your electricity supplier charges you per kWh of electricity.
4. Check the reading at different times of the day. When do you use the most energy in your home? Can you work out which appliances use the most electricity?

Just checking

1. A crane lifts a load transferring 200 000 joules of energy in 60 seconds. What is the power of the crane?

Lesson outcome

You should understand how energy transfer is measured.

1.25 Energy for everything

Link

In lesson 3.4, you will learn more about how nuclear energy is produced.

In lesson 1.26 you can learn more about renewable energy sources.

We rely on sources of energy for almost everything we do. We use energy to heat or cool homes, schools and workplaces. We use it to fuel cars, trains, airliners and other vehicles. And we use huge amounts of energy to generate electricity. Most of this energy comes from fossil fuels – oil, coal and gas. We also use nuclear power to generate electricity.

Non-renewable energy

Fossil fuels and nuclear power are **non-renewable** energy sources. Once they are used up, they cannot be replaced.

An oil platform extracting fossil fuel from deep below the seabed.

Fossil fuels

Fossil fuels are the remains of animals and plants that have been buried underground for millions of years. They are our most important sources of energy.

Energy use is increasing, but supplies of fossil fuels are limited. Experts estimate that oil supplies could run out in about 30 years. Natural gas supplies may last about 70 years, while coal reserves could last a few hundred years. These are only rough estimates, as we do not know for certain how much fuel is left in the ground.

Nuclear power stations

Nuclear power stations use energy released by splitting radioactive elements such as uranium. Supplies of uranium cannot be renewed, but they will last much longer than supplies of fossil fuels.

Nuclear power stations are expensive to build and to decommission (close down at the end of their life). Some of the wastes produced by nuclear power remain dangerous for years. Many people also worry about the effects of a nuclear accident. For these reasons, some countries do not use nuclear power.

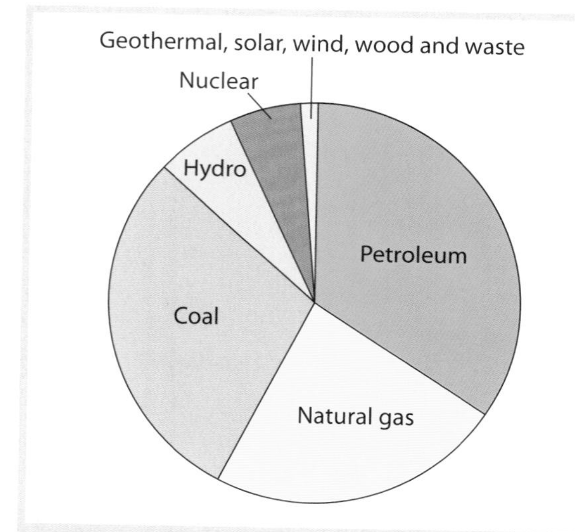

A pie chart showing the proportions of electricity generated from different fuel types in 2008. Over three-quarters of the electricity was generated from fossil fuels (petroleum, natural gas and coal), with renewable and nuclear fuels generating less than one-quarter of the electricity.

Renewable energy

Some energy sources are **renewable** – they do not run out, or they can be replaced. If we used more renewable energy we could reduce the amounts of fossil fuels we use, and so make them last longer.

Storing energy

Storing energy is important in order to balance supply and demand, and to maximise the world's sources of energy.

Most forms of renewable energy are not available all the time. For example, wind power relies on wind, and in calm periods no electricity is generated. Solar power is not generated at night or when it is cloudy. If we can store energy generated in windy periods or on sunny days, these energy resources will be more useful.

Batteries

Batteries are the main method used for storing electricity. When it is charged up, a battery is a store of chemical energy. When the battery is connected up, the chemical energy is converted to electrical energy.

Some types of battery can only be used once. Others are rechargeable. The chemical energy store can be renewed by connecting the battery to an electricity supply.

Fuel cells

Even if it is rechargeable, a battery is only a limited energy store. Once it has gone flat, it has to be recharged before it can be used again. A fuel cell is different because it can be used continuously, as long as there is fuel.

Fuel cells produce electricity using hydrogen and oxygen. They are sometimes used as a power source in space satellites and in submarines. At present fuel cells are expensive, but they could be an important type of energy store in the future.

This London bus is **hybrid**. It runs partly on an electric motor, powered by a battery that stores 75 kWh of energy.

Using energy sources effectively

The best way of ensuring that there will be energy sources for future generations is to be more efficient in our use of energy.

- In fossil fuel power stations, about 60% of the input energy is wasted in the form of heat. Only about 40% is actually transformed into electrical energy. If this wasted heat could be used to heat homes, it would save a huge amount of fuel for heating.
- A large amount of energy is released when making paper and steel, but only a fraction of this is put to use. If more of this waste energy could be used, it would save other energy supplies.

Most power stations get rid of waste heat in large cooling towers. However, CHP (combined heat and power) stations use 'waste' heat to heat local houses and buildings.

Activity A

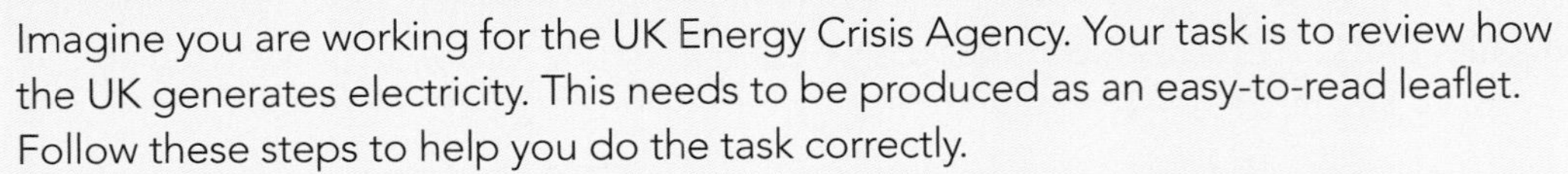

Imagine you are working for the UK Energy Crisis Agency. Your task is to review how the UK generates electricity. This needs to be produced as an easy-to-read leaflet. Follow these steps to help you do the task correctly.

1. Find out from the Internet, newspapers, local libraries, etc., how much electricity is produced by:
   (a) non-renewable sources, such as coal, natural gas, nuclear, etc.
   (b) renewable sources, such as wind, solar, hydroelectric, wave, tidal, etc.
2. Find out where these power plants are located in the UK.
3. Make a note of the advantages and disadvantages of these sources, to include cost and environmental impact.

Lesson outcomes

You should know the sources of non-renewable energy, and how energy can be stored.

1.26 Sources of renewable energy

Did you know?

The average DVD player or TV uses more electrical energy when it is on standby then when it is on, over its lifetime.

Link

You will learn more about how solar energy can be used to generate electricity in lesson 3.10.

Solar energy

Solar energy is radiation from the Sun. Solar cells are devices that can convert solar energy into electricity. They are used to power some road signs, in watches and calculators, and in space satellites. Solar energy can also be used to heat water.

Some people install solar cells on the roof to supply electricity to their home.

Sugar cane plants can be used to make bioethanol.

Wind energy

Moving air has energy, which can be used to turn wind **turbines**. The turbines turn generators, which produce electricity.

Wind turbines are fairly simple to build, so they produce electricity cheaply. However, they only work when the wind blows. Turbines are usually built in groups, in places that are windy a lot of the time. They are called wind farms. Some wind farms are built at sea.

Biofuels

Biofuels are renewable fuels made from plant materials, or from animal wastes. One type is ethanol (alcohol) made from plants. This can be used in place of petrol. In Brazil, **bioethanol** is made from sugar cane. In the USA it is made from maize (corn), and in Europe it is made from sugar beet.

Biodiesel is another kind of biofuel made from plant oils. It can be used as a fuel in diesel cars.

Biogas is made by **fermenting** manure or other kinds of waste. It can be used for heating or to produce electricity.

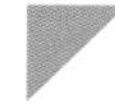

Hydroelectricity

Hydroelectric power plants generate electricity from river and rainwater held back by a dam.

The water behind the dam is a store of gravitational potential energy. When it is allowed to fall, the gravitational potential energy is transformed into kinetic energy, which can be used to turn turbines and generate electricity.

Wave energy

When wind blows across the sea, waves are formed. Large, fast-moving waves store a lot of kinetic energy.

There are several ways to use waves to generate electricity. One method is to capture the wave in a narrowing channel, where it can turn a series of turbines that generate electricity. Another method uses the up-and-down motion of waves to push air past turbines, which then turn electric generators.

The LIMPET (Land Installed Marine Power Energy Transmitter) is a wave power station on the Scottish island of Islay. It generates electricity by using the up-and-down motion of waves to push air past turbines.

Did you know?

Ocean Thermal Energy Conversion (OTEC) is a promising source of energy from the oceans. The temperature difference between the warm surface water and cold water deep below is converted into useful energy. An OTEC power plant could produce energy 24 hours a day, every day of the year. Some experimental OTEC power plants have been built.

Tidal energy

The gravitational pull of the Moon on the Earth causes the regular rise and fall of the tides. Tidal movements can be used to generate electricity.

Tides rise and fall at predictable times. This is an advantage over solar and wind energy, which are not so predictable. However, tidal energy can't be used everywhere. Tidal power stations can only be built in places where there is a large difference between high and low tides (around 5 metres).

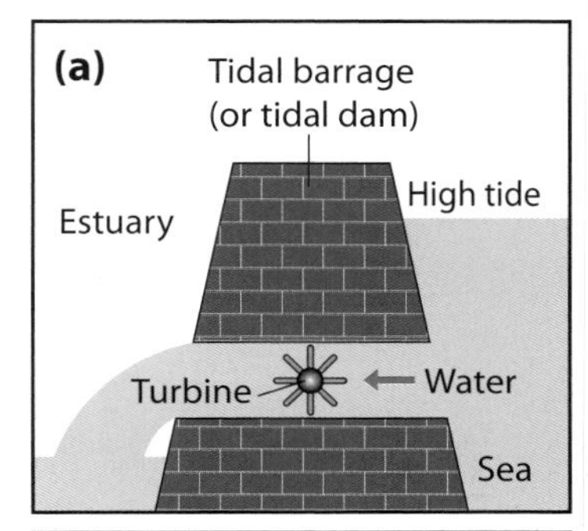

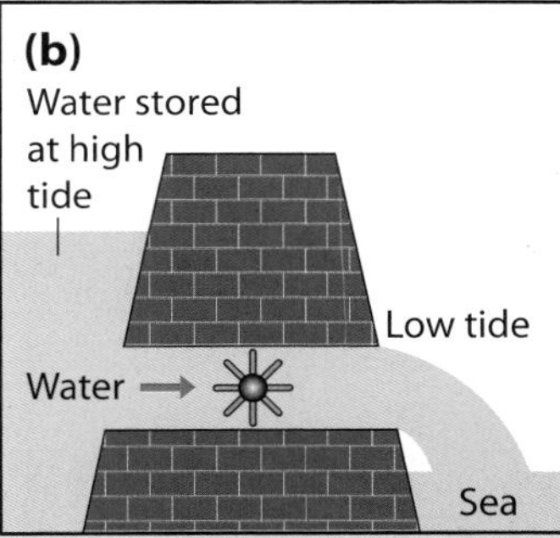

One method used to capture tidal energy is to build a barrage (barrier) across a river estuary. **(a)** At high tide, passages through the barrage are opened. Water flows through and turns turbines, which generate electricity. The passages are then closed. **(b)** At low tide, the passages through the barrage are opened. Water flows out into the sea, turning the turbines again.

Geothermal energy

Geothermal energy is heat energy generated in rocks deep underground. In some places this form of energy heats up water that then comes to the surface naturally. In other places water is pumped down boreholes drilled into underground 'hotspots'. The hot water can be used to heat buildings or to generate electricity.

Activity A

You are working for a company that is thinking of investing in renewable energy. You have been asked to find out about the different ways of generating electricity from tidal energy.

1. Use the Internet, libraries, newspapers or other information sources to find out about another way of capturing tidal energy, other than using a barrage.
2. How reliable do you see tidal energy as being for helping to meet the UK's electricity needs?
3. Compare tidal energy with wave energy and find two differences between the ways they generate electricity.

Just checking

1. Name three types of energy that are renewable.
2. For each renewable source named, describe one advantage and one disadvantage.

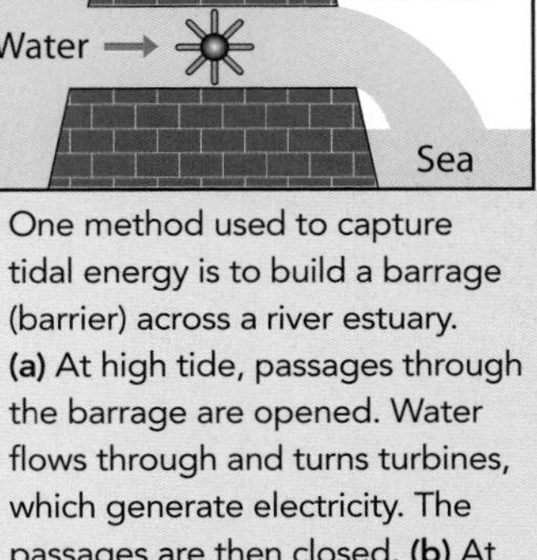

Lesson outcome

You should know the sources of renewable energy.

1.27 Wave characteristics

Key terms

Frequency – The number of waves in 1 second.

Amplitude – The maximum displacement of a wave from its rest position.

Wavelength – The length of one wave.

Discussion point

Discuss with the person next to you where you think you have used two types of waves today. What do you think are common to these waves? How would you represent these waves on a diagram?

What is a wave?

A wave is a repeated movement pattern, either up and down or back and forth, that travels from one place to another. There are many different types of waves. You can find examples all around.

A vibrating guitar string produces sound waves.

- If you drop a pebble in a pond, the ripples that spread out from the pebble are small waves in the water.
- If you pluck a guitar string or hit a drum, it creates vibrations in the air. These are sound waves. You can't see these waves, but they spread out from the sound source like the ripples from the pebble.
- Turn on a light, and light waves spread out across the room so fast that it seems to take no time at all.
- The signals that carry programmes to your TV and radio are electromagnetic waves.

wavelength
peak
amplitude
trough

Parts of a wave.

Parts of a wave

Look at the graphical representation of the parts of a wave.

- The high parts of the wave are called peaks and the low parts are called troughs.
- The distance between two successive peaks or troughs is called a **wavelength** and is normally measured in metres.
- The **amplitude** of a wave represents the maximum displacement from the centre line.
- The number of waves that pass a point in 1 second is called the **frequency**. The unit of frequency is the hertz (Hz).

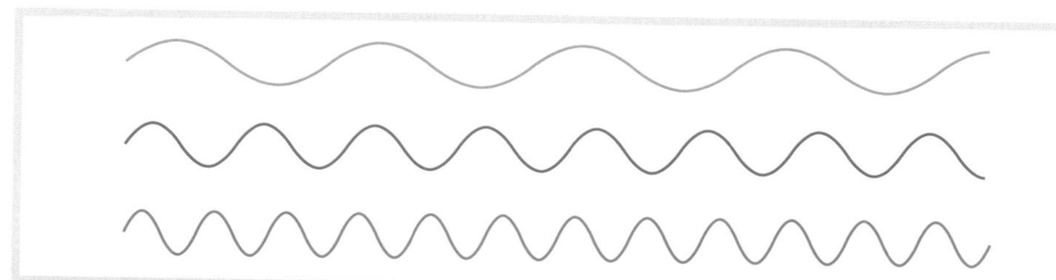

Graphical representation of three waves. The wave with the highest frequency is the pink one – as it repeats itself the most within the same timescale. The wave with the lowest frequency is the orange one.

The speed of a wave

The **speed** of a wave is related to frequency and wavelength by this equation:

wave speed = wavelength (m) × frequency (Hz)

The unit of wave speed is metres per second (m/s).

Worked example

A wave has a wavelength of 5 m and frequency of 1000 Hz. What is the speed of the wave?

Step 1 Write down the equation: speed of wave = wavelength × frequency.

Step 2 Substitute the numbers into the equation:
speed of wave = 5 × 1000 = 5000 m/s.

A wave has a frequency of 10 MHz and speed of 300 000 000 m/s. What is the wavelength of the wave? Use standard form for your answer.

Step 1 Change to standard form: 10 MHz = 10^7 Hz; 300 000 000 m/s = 3×10^8 m/s.

Step 2 Write down the equation: speed of wave = wavelength × frequency.

Step 3 Make the wavelength the subject: wavelength = speed of wave ÷ frequency.

Step 4 Substitute the numbers into the equation:
wavelength = $3 \times 10^8 \div 10^7$ = 30 m. **Hint**: don't forget the units.

Remember

In science we use standard form to write large or small numbers. Light waves have very high frequencies and very short wavelengths, and standard form is often the best way of writing these down.

For example, a light wave may have a wavelength of 0.0000007 m and a wavelength of 430 000 000 000 000 Hz. This is much better written as 7×10^{-7} m and 4.3×10^{14} Hz.

Case study

Anosh is a technician working for a TV broadcaster. His job is to make sure that the video transmissions they send out are sharp, and the sound is crystal clear. Anosh is trained in electronics, computing and broadcast technology. He uses special equipment to monitor the TV signals. For example, he checks that the picture signal has the correct brightness and contrast, and that the audio signal is strong enough.

1 What kind of waves is Anosh working with?

2 What speed do these waves travel at?

Link

You will learn about speed of an electromagnetic wave in lesson 1.28.

Activity A

Select two of your favourite radio stations, one local and one national.

1 Write down the frequencies of these two radio stations.

2 Work out the wavelength of the radio stations. **Hint**: What kinds of waves are these? If you know the type of wave you can find out its speed. You need this to work out the wavelength.

Just checking

1 The wave traces shown on page 66 were recorded over 1 second. What is the frequency of the pink wave?

2 The speed of the pink wave is 60 m/s. What is its wavelength?

Lesson outcomes

You should know the characteristics of a wave, how to use the wave speed equation and how to express values in standard form.

Get started

Discuss in groups the features of a rainbow. When you see a rainbow in the sky, how are the colours arranged? Why do you think these colours are arranged in this way?

Key term

Electromagnetic wave – A wave made up of electric and magnetic fields – for example radio waves, infrared radiation, visible light and X-rays.

Did you know?

Light takes 8 minutes to reach the Earth from the Sun.

The electromagnetic spectrum

If you study a rainbow or the spectrum from a prism, you will notice that the colours are always in the same order: red, orange, yellow, green, blue, indigo, violet. The different colours are different wavelengths of light. Red has the longest wavelength, (lowest frequency), violet the shortest wavelength (highest frequency). Light is one form of **electromagnetic radiation**. The spectrum of light that we see in a rainbow is just part of a much larger electromagnetic spectrum.

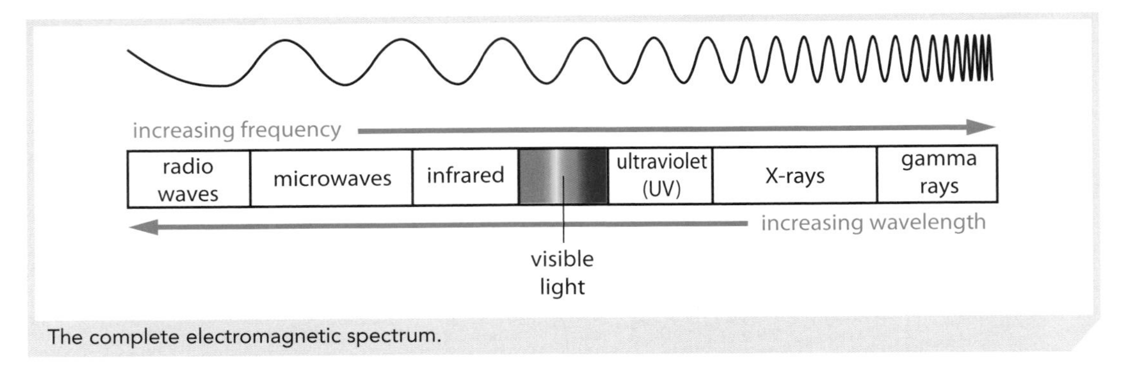

The complete electromagnetic spectrum.

The electromagnetic spectrum is continuous, but we split it into sections according to the properties of the waves. At one end are radio waves, which can have wavelengths of 100 km. At the other are gamma rays, with wavelengths as small as 10^{-12} m.

The table gives more detailed information about the different parts of the electromagnetic spectrum, from radio waves to gamma rays. The table lists some uses of each type of wave, and possible harmful effects.

	Radio waves	Microwaves	Infrared
Typical wavelength	10–200 m	1 mm–1 cm	10^{-5} m
Uses	Broadcasting Satellite transmissions	Satellite transmissions Microwave ovens Communications Weather forecasting	Cooking (ovens, grills and toasters) Thermal imaging Optical fibres TV remote control
Harmful effects		Internal heating of body cells	Skin burns

Radio broadcasts carried by radio waves.

A microwave oven.

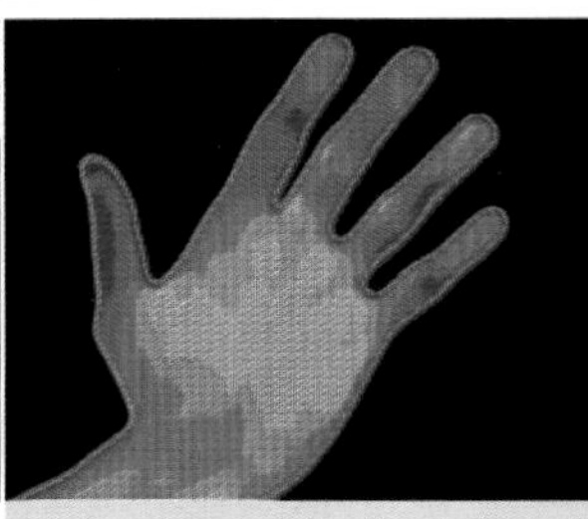

A hand viewed using infrared.

Properties of electromagnetic waves

Waves with a long wavelength and low frequency have less energy than waves with a short wavelength and high frequency. High-energy forms of electromagnetic radiation have more harmful effects on humans.

The different types of electromagnetic radiation have some things in common. All **electromagnetic waves** travel at the same speed – the speed of light. In a vacuum or in air, this speed is 300 million metres per second (3×10^8 m/s). All electromagnetic waves can travel through a vacuum. This is different from waves such as sound or waves in water, which need some kind of medium for the wave to travel in.

Activity A

Look at the section of the table below on harmful effects of electromagnetic radiation. Some kinds of electromagnetic radiation can be harmful, but we still manage to use them safely.

Find two or three safety precautions that are used to protect people from the harmful effects of electromagnetic radiation.

Just checking

1 Where does the light that falls on the Earth come from?
2 What is the frequency of the radiation that is transmitted by a TV remote control?
3 The order of the colours of the visible spectrum can be remembered by forming a boy's name: ROY G. BIV. Can you think of a way of remembering the order of the sections of the electromagnetic spectrum?

Lesson outcomes

You should know what types of radiation comprise the electromagnetic spectrum, and their uses and harmful effects.

Visible light	Ultraviolet	X-rays	Gamma rays
5×10^{-7} m	10^{-8} m	10^{-10} m	10^{-12} m
Display lights Photographs	Fluorescent lamps Detecting forged bank notes Disinfecting water	Observing the internal structure of objects Medical X-rays	Sterilising food and medical equipment Detection of cancer and its treatment
	Damage to surface cells and eyes, leading to skin cancer and eye conditions	Mutation or damage to cells in the body	Mutation or damage to cells in the body

Visible light.

UV scanning of bank notes to check for forgeries.

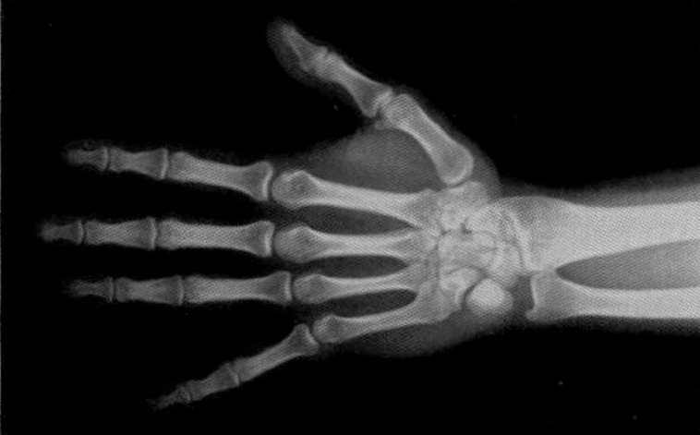

An X-ray image of a hand.

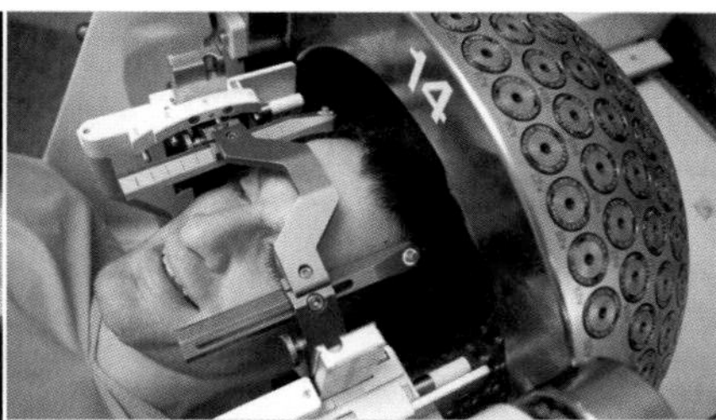

Gamma rays being used to treat a brain tumour.

Assessment Zone

How you will be assessed

You will take a written test on the same day and time as other learners – your teacher will let you know the date of the test. The test will last one hour and you should aim to answer all the questions on the paper. For some questions you will need to use a calculator.

Your BTEC written test will contain these types of questions:

- **multiple-choice questions** where the answers are available and you have to choose the answer(s) that fit(s).
- **short-answer questions** where you are asked to give a short answer worth 1–2 marks.
- **long-answer questions** where you are asked to give a longer answer which could be worth up to 6 marks.

Test tips

- At the start make sure you have read the instructions, that you can see the clock and that you are comfortable to write.
- Watch the time – you should aim to spend about a 'minute per mark'. Some early questions will take less time than this and some later ones more.
- If you get stuck on a question move onto the next one and come back to that question at the end.
- The space given for your answer will show you the type of answer required, e.g. if two answers are required you may see the answer space divided up for two answers.
- Remember that you can use more paper if necessary, e.g. because you have made a mistake or you need more space for your answer.
- Plan your longer answers – read the question carefully and think about the key points you will make.

Please note: the questions on the following pages are for practice. They do not represent a real test paper.

Assessment practice

Here are some questions for each of the three sciences: biology, chemistry and physics. Tips are given to help you understand what the question is asking and how to answer it. Full answers to the questions have also been included.

Biology

Question 1

(a) The table shows five different structures found in cells. Show whether each structure is found in animal cells and/or plant cells. Put a tick (✓) or a cross (✗) in the correct box(es) for each cell. The first one has been done for you. [2]

Structure	Plant cell	Animal cell
Cell membrane	✓	✓
Chloroplasts	✓	✗
Mitochondria	✓	~~✗~~ ✓
Cell wall	✓	✗
Vacuole	✓	✗

If you get one or two wrong you would get one mark. Three or four wrong and you get 0 marks.

If you change your mind, always cross out completely the one you want to change and insert a new tick or cross, so that it is clear whether you have used a tick or a cross.

This candidate has got all four rows correct and so gets two marks.

(b) Arrange the following in order of size.
Start with the smallest and end with the largest. [1]

nucleus	***chromosome***	***gene***	***cell***	***organ***	***organism***

Smallest Largest

gene chromosome nucleus cell organ organism

This is a correct answer and gets the mark. There is only one mark so they must ALL be correct.

Question 2

(a) Which one of the following describes the term allele?
Tick the correct answer. [1]

A dominant gene	☐
A version of a gene	✓
A length of DNA	☐
How nucleotide bases pair	☐

This is a key term and you need to learn the definitions for all key terms.

(b) Ella and James run in a 400 metre race. At the end of the race they both feel hot because their body temperature has risen. Tick all of the statements on page 72 that correctly describe how their bodies respond to this rise in temperature. [2]

This question is asking you to apply what you know about temperature regulation to a particular situation. You need to think about how your body loses excess heat.

The question indicates that more than one of the statements may be correct. Statements 2, 3 and 4 describe what happens when your body temperature drops, so are not correct here.

Your nervous system and skin help bring your body temperature back to normal	✓
You sweat less and lose less heat by evaporation	☐
The hairs on your skin stand up	☐
Blood vessels in your skin undergo vasoconstriction	☐
Regulating your body temperature is an example of homeostasis	✓

This question tests what you can work out from the numbers in this symbol. Make sure that you know the meanings of mass number and atomic number and which number in the symbol is which.

Chemistry

Question 1

Here is the symbol for an atom of bromine-81: $^{81}_{35}Br$

(a) Look at the table below.
Put a tick next to the correct statement. [1]

The atomic number of bromine is 81	☐
This atom has 35 neutrons	☐
The total number of neutrons and protons in the nucleus of this atom is 81	✓
The mass number of this atom is 46	☐

(b) There is another isotope of bromine, bromine-79, with the symbol $^{79}_{35}Br$.

$$^{81}_{35}Br \qquad ^{79}_{35}Br$$

Compare the two isotopes in terms of their atomic number and mass number. [1]

Explain what this means in terms of protons and neutrons. [1]

The two isotopes have the same atomic number but different mass numbers.

They have the same number of protons (35) but different numbers of neutrons. One has 44 neutrons and the other has 46 neutrons.

Knowing the definition of the word isotope will help here, but you should also show that you have used the information given in the two symbols.

(c) The relative atomic mass of an element can be worked out from the masses and abundances of the different isotopes.

A sample of 100 bromine atoms contains 50 atoms of bromine-79 and 50 atoms of bromine-81.

Calculate the relative atomic mass of bromine. Show your working in the box. [2]

Mass of bromine-79 atoms = 50 × 79 = 3950

Mass of bromine-81 atoms = 50 × 81 = 4050

Total mass of 100 atoms = 8000

Average mass = $\frac{8000}{100}$ = 80

The working shows that you can use the information you have been given.

You would gain 1 mark for the first two calculations and 1 mark for the second two calculations.

Even if your answer to the first part is wrong, you may still get 1 mark in the second part for using the correct method.

Question 2

The diagram below is a simplified version of the periodic table, showing the positions of five different elements.
Look at the diagram and use the chemical symbols in it to answer the questions that follow.

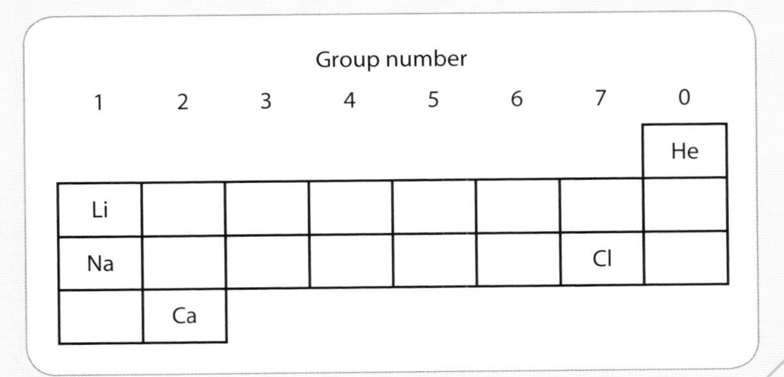

(a) Which two elements are in the same period? [1]

Na and Cl

Periodic tables often show you group numbers, but you need to know the difference between groups and periods.

(b) Which element only has two electrons in its outer shell? [1]

Ca

(c) Which two elements are non-metals? [1]

Cl and He

(d) Lithium and sodium have similar chemical properties.
Explain how you can tell this from the periodic table. [2]

Because they are in the same group which means that they will have the same number of electrons in their outer shell.

There are two marks for this question, so you need to make sure that you make two different points, but your answer needs to refer to the periodic table.

When you are asked to describe a test you should always give the result you expect as well as describing how to do the test.

Often, only one or two numbers need to be put in front of parts of an equation to make it balance.

Question 3

Dilute hydrochloric acid reacts with zinc. One of the products of this reaction is a gas.

(a) Jamie thinks the gas could be hydrogen. Describe how he could carry out a test to show this. [2]

Collect a test-tube of the gas and then try and light it with a lighted splint. If it is hydrogen, it will burn with a squeaky pop.

(b) The other product of the reaction is a salt. Name the salt. [1]

Zinc chloride

(c) In another reaction, Jamie uses calcium instead of zinc. Balance this chemical equation for the reaction. [1]

$Ca + 2HCl \rightarrow CaCl_2 + H_2$

Physics

Remember the time in this equation is in hours, so if you are given a time in minutes you have to convert to hours by dividing the number by 60.

The number has been rounded up.

Question 1

Adham wanted to calculate the cost of electricity he and his friends used while watching the FA Cup Final.

He knows that the equation to use is:
energy transferred (kWh) = power (kW) × time (h)

During the football match the television was switched on. It has a power rating of 0.4 kW. They spent 3.5 hours watching the match.

(a) How much energy is transferred? [1]

energy transferred (kWh) = power (kW) × time (h)
energy = 0.4 × 3.5 = 1.4 kWh

(b) The electricity company charges 13.5p per kWh.
How much did it cost them to have the television on? [1]

cost = 1.4 × 13.5 = 18.9p = 19.

The example shows that the learner fully understands the how hydroelectric energy is used to generate electricity. This answer would gain full marks.

(c) Hydroelectric power is a renewable source of energy. Explain how it is generated. [4]

River water and rain water is collected and stored in a reservoir behind a dam.

The water in the reservoir is storing gravitational potential energy.

When the water is then allowed to fall through turbines, the gravitational potential energy is then being transformed into kinetic energy which is then used to generate electricity.

Question 2

(a) The table below shows the electromagnetic spectrum.
Fill in the missing types of wave. [2]

Radio waves
Infrared
Ultraviolet

Radio waves
Microwaves
Infrared
Visible light
Ultraviolet
X-rays
Gamma rays

In this question you need to make sure that you know how to use standard form. Practice standard form before the test.

(b) An X-ray machine transmits X-rays that have a frequency of 2×10^{17} Hz and travel at 3×10^8 m/s in air.
The speed of the wave is given by the equation:

wave speed = wavelength × frequency

Calculate the wavelength of the X-ray, showing your working. [3]

Always transpose to find the subject of the equation (in this case the wavelength) before you substitute the numbers. Don't forget the units.

wave speed = wavelength × frequency

wavelength = wave speed/frequency

$$\text{wavelength} = \frac{3 \times 10^8}{2 \times 10^{17}} = 1.5 \times 10^{-9}\,\text{m}$$

How to improve your answers

Here is a question that assesses the chemistry you will have learnt as part of this unit.

Describe what neutralisation is, including reactants and products, and how it can be used to treat an environmental problem. [6]

Read the three learner answers below, together with the feedback.

This activity will help you learn more about how to answer longer answer questions. Try to use what you learn here when you answer questions like this in your test.

The word neutralise isn't actually described.

Why this is important is not described.

Learner 1

Neutralisation means to neutralise an acid. When lakes are acid they can be neutralised.

Feedback:

'This answer might pick up 1 or 2 marks because the word 'acid' is mentioned in the first sentence, and an example of an important neutralisation reaction is given. But the candidate hasn't described what 'neutralise' actually means, or how the lake is neutralised.

There is no mention of what happens when a base and an acid react together.

The actual substance used isn't named.

Learner 2

In a neutralisation reaction a base reacts with an acid. This is used when lakes are acidic because of acid rain. It is added to the lake and this makes the environment better for fish, etc.

Feedback:

This would probably pick up 3 or 4 of the 6 marks available. The candidate has made one important point about neutralisation but the description of the reaction is still not complete. They have then applied this information to describe how this would be used to treat an acidic lake, but have given very little extra information about how this would be done.

A full description of neutralisation is given.

The environmental problem is described.

Details are given about how the lake is treated, and what will happen after the treatment.

Learner 3

Neutralisation is when a substance like a base or alkali is added to an acid. A salt and water are produced and the pH becomes neutral (pH = 7).

Some lakes in Scandinavia have become acidic (pH = 3 or 4) because of acid rain. Calcium carbonate powder is added to the lake. This neutralises the acid and the pH returns to normal (around pH 6.5) so that fish can live and breed in the lake.

A base is used instead of an alkali because a base would only neutralise the acid and would not then continue to react, as it is not soluble in water. An alkali is soluble in water and could make the lake alkaline. This would cause similar problems to the lake being acidic.

Feedback:

This is an excellent answer that covers all the required points and will score 5 or 6 marks. Neutralisation is described very well and the candidate uses ideas about pH to develop their answer even more. This is very helpful for the second part of the answer where the candidate gives details about the pH of the acidic lakes. They also describe clearly how the neutralisation is carried out.

Test tips for this question

- You need to know the full meaning of scientific words, such as 'neutralisation'.
- If there are several marks available for an answer, then you will need to give as much detail as possible when you try and explain the meaning of these words.
- When a question asks you to **describe** a process, you will also need to make several different points and often organise them into the correct order.
- In this question you had to *identify* an environmental problem, *describe* the problem and then say how neutralisation can *help to treat* the problem.
- Try and avoid using words such as 'it' or 'they' in your answers because it may not be clear which word you actually mean.
- Make sure that you use correct **scientific words** in your answers, like the full names of any chemical substances or, in this case, the term 'pH'.

Assess yourself

Biology

Question 1

(a) Choose functions from the list below and write the correct letter, A–E, for each structure. [2]

Structure	Function
Leaf	
Root	
Sensory neurone	

Functions

A Anchors the plant in soil and takes up water

B Pumps blood around the body

C Carries impulses from the central nervous system to muscles or glands

D Carries out photosynthesis

E Carries impulses from receptors to the central nervous system

(b) The diagram shows the structure of an animal cell.

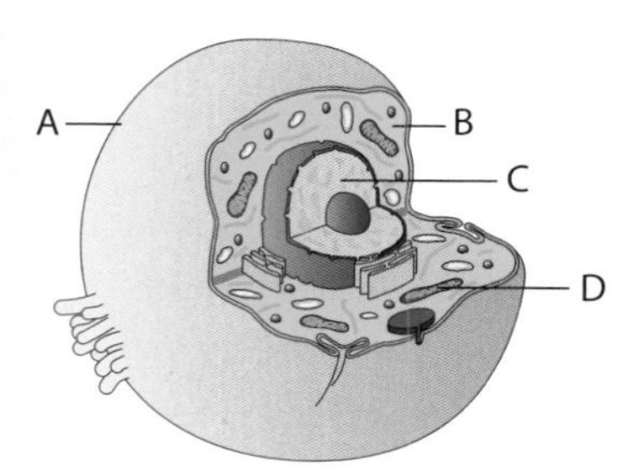

(i) Write the names of the parts indicated by the label lines A–D. [2]

(ii) State the function of part B. [1]

(iii) Describe one way that the structure of a plant cell would differ from this animal cell. [1]

Question 2

(a) Two parents are heterozygous for a gene that controls the characteristic hair colour.

Fair hair is recessive and the symbol for its allele is b.

Dark hair is dominant and the symbol for its allele is B.

Calculate the probability of their child having fair hair.

Show your working. [3]

(b) Explain how genetic mutations (changes to genetic material) can change the characteristics of an organism. [2]

Question 3

Which one of the following describes homeostasis? Tick the correct answer. [1]

Sweating when you are hot so heat is lost by evaporation ☐

Chemicals carrying information across synapses ☐

Hormones travelling in the blood to target organs ☐

Keeping the internal conditions of the body constant ☐

Genetic mutations that are beneficial ☐

Question 4

(a) How may the blood glucose level change about an hour after eating a meal? [1]

(b) Explain how the body responds to the change in blood glucose level after eating. [3]

Question 5

Write a number between 1 and 5 in each box to show the correct order to describe the pathway taken in a reflex arc:

	Impulse travels to effector which carries out an action, e.g. muscle contracts
	Nerve impulse travels along sensory neurone
	Receptor in muscle detects a change in environment
	Impulse travels along motor neurone
	Impulse travels to a motor neurone via a synapse in the spinal cord

[2]

Total: [18]

Chemistry

Question 1

The diagram below shows the structure of a potassium isotope. Potassium has an atomic number of 19.

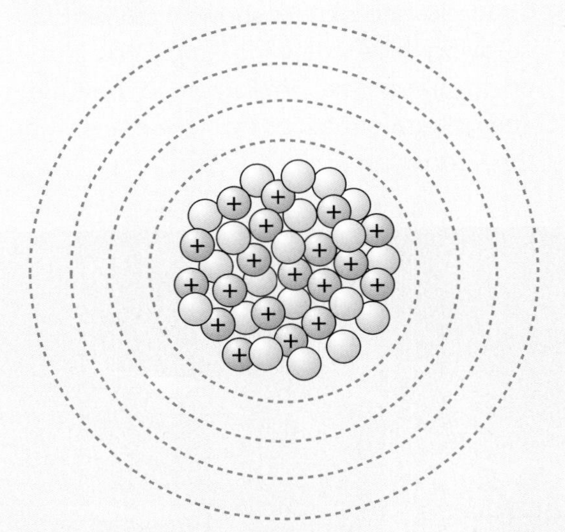

(a) Complete the diagram to show the electron configuration of the potassium atom. [1]

(b) The symbol for this atom of potassium is written $^{39}_{19}K$. A second isotope of potassium has two extra neutrons. Write the symbol for this isotope. [1]

Question 2

Below are four statements about the nucleus of atoms. Put a tick in the box next to the correct statement. [1]

The nucleus always contains equal numbers of protons and neutrons	☐
The nucleus has a negative charge	☐
The nucleus takes up much less space than the rest of the atom	☐
The nucleus contains protons, neutrons and electrons	☐

Question 3

Poppy is investigating the properties of sulfuric acid.

(a) She adds a few drops of litmus solution to a solution of sulfuric acid. What colour does she observe? [1]

(b) She does a second experiment. She adds solid copper carbonate to sulfuric acid. The solid dissolves to form a blue solution of a salt.

(i) What is the name of the salt in the blue solution? [1]

(ii) Explain what else Poppy observes. [2]

(c) Poppy also investigates how calcium oxide can neutralise sulfuric acid.

(i) Describe one way in which calcium oxide is useful in farming. [1]

(ii) When calcium oxide, CaO, neutralises sulfuric acid, H_2SO_4, one of the products is calcium sulfate, $CaSO_4$. Write the balanced chemical equation for this reaction. [1]

Question 4

Richie has a sample of a white solid which could be calcium oxide or calcium carbonate.

To decide which substance it is, he adds hydrochloric acid to the solid. He tests the gas given off to see if it is carbon dioxide. The test is positive.

(a) Describe how to carry out a test for carbon dioxide, and what Richie will see if the test is positive. [3]

(b) (i) What can Richie conclude is the name of the white solid? Explain your answer. [1]

(ii) Give the formula of the white solid. [1]

Question 5

You have two different containers labelled A and B.

- Container A contains air.
- Container B contains water.

(a) Complete the table showing whether each substance is an element, a compound or a mixture.

Container	Contains	Element, compound or mixture?
A	Air	
B	Water	

[2]

(b) Explain how you arrived at your answers. Include descriptions of elements, compounds and mixtures, where appropriate, in your answer. [2]

Total: [18]

Physics

Question 1

(a) Pictures (i), (ii) and (iii) show three devices that convert energy into different forms. Name the useful form of energy for each device. [3]

(i)

(ii)

(iii)

(b) The following picture shows a student studying under a reading lamp. Draw an energy transformation diagram to show the energy transformations that take place when the lamp is on. [2]

(c) The input energy to the lamp is 200 J and the useful energy transferred into light energy is 50 J.

(i) Calculate how much energy is wasted. [1]

(ii) Calculate the efficiency of the lamp. [1]

(b) BBC Radio 1 transmits radio waves of frequency 97.7 MHz. If the speed of the waves is 3×10^8 m/s, calculate the wavelength of the waves using the equation:

speed of wave = frequency × wavelength [2]

Question 2

(a) Which of the following is a renewable source of energy?
Tick **one** of these boxes: [1]

coal ☐

wind ☐

nuclear ☐

oil ☐

(b) Give **two** advantages and **two** disadvantages of renewable energy over non-renewable energy sources. [4]

Question 3

(a) The table below shows some of the uses and some of the harmful effects of types of electromagnetic waves. Complete the rest of the table.

Type of electromagnetic wave	Use	Harmful effect
Microwaves	Cooking	Internal heating of body cells
Infrared		
Ultraviolet		Skin cancer

[3]

Question 4

Complete the table below. Add convection, conduction or radiation beside each statement.

Type of thermal energy transfer	Description
	Occurs best in solid materials.
	Can travel through a vacuum and does not need particles for energy transfer to occur.
	Occurs in liquids or gases.
	Houses in hot countries are painted white to minimise this type of energy transfer.
	A radiator uses this type of thermal energy transfer to warm up a room.
	Handles of cooking pans are usually wooden to minimise this type of energy transfer.

[3]

Total: [18]

Introduction

A glass of water, the oxygen in the air, the silicon chips in your mobile phone – all of them are chemicals and all of them are made of just one or two different kinds of atom. But each chemical has very different properties and is used in different ways in our daily lives.

In this unit you will see how atoms can bond together in different ways and how that helps to explain their properties. You will study how some chemical reactions happen and look for patterns in the reactions.

You will also study how chemistry is used in industry to produce important new substances, and use practical investigations to discover how these reactions can be made quicker and cleaner.

The Earth has changed a lot over its history. You will find out more about this. You will also consider what choices society may have to make to prevent the Earth changing in ways that could damage the environment for future generations.

Assessment: You will be assessed using a series of internally assessed assignments.

Learning aims

After completing this unit you will:

a have investigated chemical reactivity and bonding

b have investigated how the uses of chemical substances depend on their chemical and physical properties

c have investigated the factors involved in the rate of chemical reactions

d understand the factors that are affecting the Earth and its environment.

I've enjoyed all the experiments we have done in this unit. Our teacher showed us some amazing reactions like when she added potassium to water.

Thinking about pollution and global warming is a bit scary but it was good to find out what scientists are doing to try and tackle these problems.

Kwabena, *15 years old*

Chemistry and Our Earth

2

This table shows you what you must do to achieve a Pass, Merit or Distinction grade, and where you can find activities in this book to help you.

Assessment criteria			
To achieve a Level 1 Pass grade, the evidence must show that you are able to:	To achieve a Level 2 Pass grade, the evidence must show that you are able to:	To achieve a Level 2 Merit grade, the evidence must show that you are able to:	To achieve a Level 2 Distinction grade, the evidence must show that you are able to:
Learning aim A: Investigate chemical reactivity and bonding			
1A.1 Classify group 1 and 7 elements based on their physical properties. Assessment activity 2.2	**2A.P1** Describe the physical and chemical properties of group 1 and 7 elements. Assessment activity 2.2	**2A.M1** Describe trends in the physical and chemical properties of group 1 and 7 elements. Assessment activity 2.2	**2A.D1** Explain the trends in chemical properties of group 1 and 7 elements in terms of electronic structure. Assessment activity 2.2
1A.2 Describe properties of ionic and covalent substances. Assessment activity 2.1	**2A.P2** Compare properties of ionic and covalent substances. Assessment activity 2.1	**2A.M2** Explain the properties of ionic and covalent substances. Assessment activity 2.1	**2A.D2** Relate applications of compounds to their properties and to their bonding and structure. Assessment activity 2.1
1A.3 Classify substances as ionic or covalent. Assessment activity 2.1	**2A.P3** Draw dot-and-cross diagrams of simple ionic and covalent substances. Assessment activity 2.1	**2A.M3** Describe the formation of ionic and covalent substances. Assessment activity 2.1	
Learning aim B: Investigate how the uses of chemical substances depend on their chemical and physical properties			
1B.4 Describe physical properties of chemical substances. Assessment activity 2.3 Assessment activity 2.4	**2B.P4** Describe how chemical substances are used based on their physical properties. Assessment activity 2.3 Assessment activity 2.4	**2B.M4** Explain how physical and chemical properties of chemical substances make them suitable for their uses. Assessment activity 2.3 (part) Assessment activity 2.4 (part)	**2B.D3** Assess the suitability of different types of substance for a specified use. Assessment activity 2.3 (part) Assessment activity 2.4 (part)
1B.5 Describe chemical properties of chemical substances. Assessment activity 2.4	**2B.P5** Describe how chemical substances are used based on their chemical properties. Assessment activity 2.4		

Assessment criteria			
Learning aim C: Investigate the factors involved in the rate of chemical reactions			
1C.6 Identify the factors that can affect the rates of chemical reactions. Assessment activity 2.5	**2C.P6** Describe the factors that can affect the rates of chemical reactions. Assessment activity 2.5	**2C.M5** Explain how different factors affect the rate of industrial reactions. Assessment activity 2.5	**2C.D4** Analyse how different factors affect the rate and yield of industrial reactions. Assessment activity 2.6
1C.7 Identify reactants and products, including state symbols in chemical equations, and whether reactions are reversible or irreversible. Assessment activity 2.5	**2C.P7** Identify the number and types of atoms in balanced chemical equations. Assessment activity 2.5	**2C.M6** Explain the terms 'yield' and 'atom economy' in relation to specific chemical reactions. Assessment activity 2.6	
Learning aim D: Understand the factors that are affecting the Earth and its environment			
1D.8 Identify the human activities that affect the Earth and its environment. Assessment activity 2.8	**2D.P8** Describe the human activities that affect the Earth and its environment. Assessment activity 2.8	**2D.M7** Discuss the extent to which human activity has changed the environment in comparison to natural activity. Assessment activity 2.7 (part) Assessment activity 2.8	**2D.D5** Evaluate possible solutions to changes in the environment, occurring from natural or human activity. Assessment activity 2.8
1D.9 Identify natural factors that have changed the surface and atmosphere of the Earth. Assessment activity 2.7	**2D.P9** Describe natural factors that have changed the surface and atmosphere of the Earth. Assessment activity 2.7		

How you will be assessed

The unit will be assessed by a series of internally assessed tasks. You will be expected to show an understanding of chemistry relevant to industrial processes, environmental issues and to the use of chemicals across a wide range of industries.

The tasks will be based on scenarios which place you as the learner in the position of working in a number of industrial and environmental sectors, for example, in the research and development department of a chemical firm, as a laboratory technician or as an environmental scientist.

Your actual assessment could be in the form of:

- written materials such as a training manual or report
- a log of experimental observations and results
- a presentation or DVD describing important scientific principles.

2.1 Types of bonding

Get started

You will have been told that everything is made of atoms, but you have never seen one! So what do we know about atoms? In groups, try and think of as many facts about atoms as possible.

Key terms

Molecule – Two or more atoms held together by covalent bonds.

Ion – An atom (or group of atoms) with a positive or negative charge.

Bonding – The way in which particles are held together in chemical substances.

Structure – The way in which particles are arranged in chemical substances.

Formula (plural formulae or formulas) – A way of showing the number of atoms of each type which bond together.

Ionic bonding – Bonding between positive and negative ions.

Covalent bonding – Bonding between atoms which are sharing one or more pairs of electrons.

Link

Lessons 1.10 and 1.11 explain the structure of atoms.

The variety of chemical substances

Scientists have made and studied millions of different substances. All those substances have different properties – why? The answer comes from knowing about the particles that make up the substances and:

- how they are held together (the **bonding**)
- how they are arranged (the **structure**).

Covalent bonding and molecules

Some of the simplest and commonest substances in the universe, like hydrogen gas and water, are held together by **covalent bonds**. These substances are made up of **molecules**.

As you can see from the diagram, hydrogen gas contains molecules in which two hydrogen atoms are bonded together. Chemists write the **formula** of hydrogen gas as H_2 to show this. Each hydrogen atom has one electron in a shell outside the nucleus. It shares this electron with the other atom and the negative charge of the electrons attracts the positive nucleus in each atom. This bonds the atoms together.

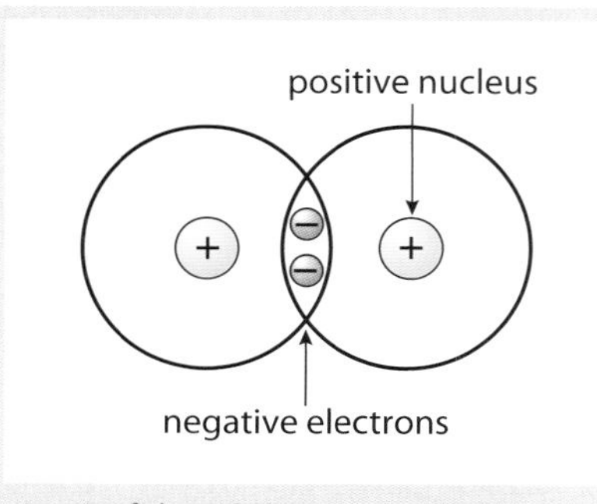

A pair of shared electrons holds two hydrogen atoms together.

When atoms share electrons to form molecules, each atom often ends up with a full outer shell. This helps to make the molecules stable.

Some molecules, such as methane or water, may contain two or more different types of atom. Look at the diagrams below which show you how the atoms and the electrons are arranged in methane and water molecules.

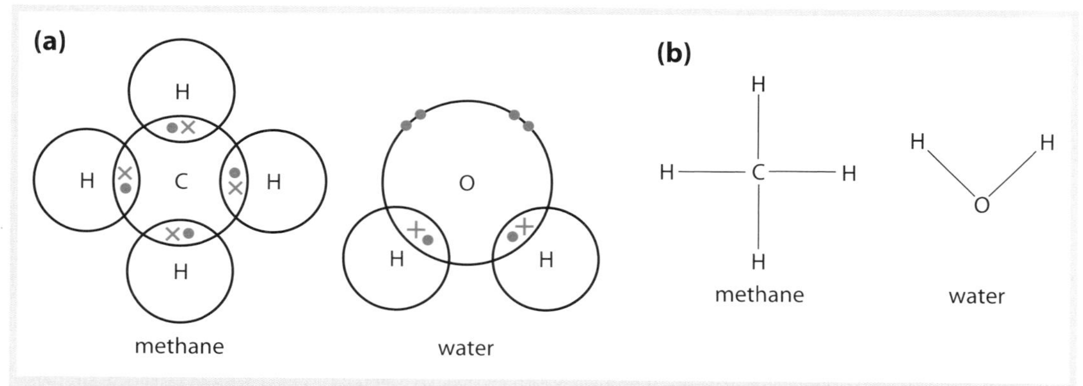

(a) The electron arrangement of methane and water, **(b)** the structure of these molecules. Each line in the diagrams shows a single covalent bond where a pair of electrons is being shared.

Cl—Cl chlorine | O=C=O carbon dioxide | O=O oxygen

The structure of molecules of chlorine, carbon dioxide and oxygen. A double line shows you that the molecule contains a double covalent bond, where two pairs of electrons are being shared.

Ionic bonding

Sharing electrons isn't the only way for atoms to bond together. When sodium (a metal) reacts with chlorine (a non-metal) an ionic compound, sodium chloride, is formed. Each sodium atom loses an electron to become a positively charged sodium **ion**. Each chlorine atom gains an electron to become a negatively charged chloride ion.

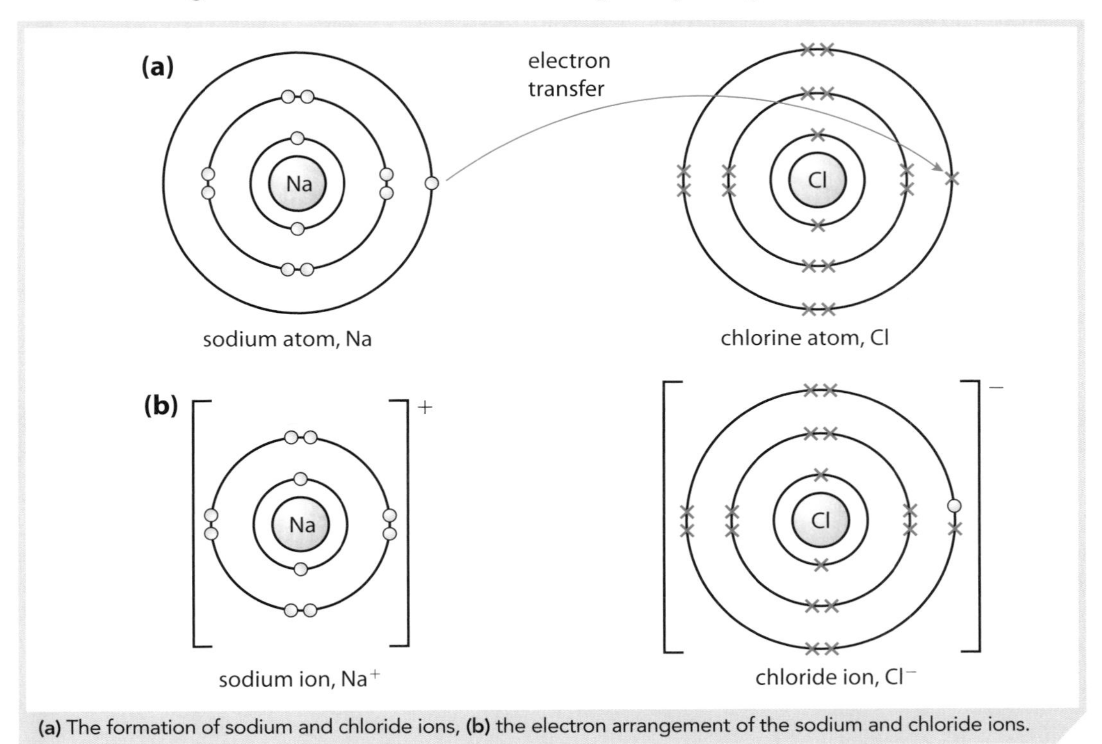

(a) The formation of sodium and chloride ions, **(b)** the electron arrangement of the sodium and chloride ions.

There are equal numbers of sodium and chloride ions in the compound so the formula is written as NaCl.

Other common ionic compounds are magnesium chloride ($MgCl_2$) and magnesium oxide (MgO).

Activity B

Magnesium has two electrons in its outer shell and oxygen has six. When magnesium reacts with oxygen, magnesium loses two electrons to form a magnesium ion and oxygen gains two electrons to form an oxide ion.

1. Draw diagrams to show the formation of magnesium and oxide ions and the electron arrangement of the magnesium and oxide ions.
2. The formula of magnesium oxide is MgO. Explain why.

Just checking

Copy and complete this sentence about bonding:

Substances can be held together by covalent bonds where electrons are _______ between atoms, or by _________ bonds where electrons are lost or gained by atoms.

Activity A

1. Use the information in the diagrams on this page to work out the formulae of these molecules: **(a)** methane, **(b)** water, **(c)** chlorine, **(d)** carbon dioxide and **(e)** oxygen.
2. Draw diagrams of the electron arrangements for **(c)** to **(e)** – or look them up in books or on websites.

Remember

When two non-metal elements react together the bonding in the compound will be covalent.

If a non-metal reacts with a metal the bonding in the compound will be ionic.

Take it further

There are no small molecules in ionic compounds, so the formula of an ionic compound shows the ratio of each type of atom.

Magnesium chloride has the formula $MgCl_2$ so there are twice as many chloride ions as magnesium ions.

Lesson outcomes

You should:
- know what is meant by the formulae of molecules
- understand covalent and ionic bonding.

2.2 Properties of substances

Get started

Sodium chloride (table salt), candle wax, and sand are three familiar substances in everyday life. Think about these substances and list as many physical properties as you can for each substance.

Take it further

Forces between molecules are called intermolecular bonds.

Link

You will read more about the connection between properties and uses in lessons 2.5 and 2.6.

Ionic substances

When positive and negative ions form, they are very strongly attracted to each other. But even after two ions have bonded together they still attract other ions to them. So ionic compounds don't form small molecules – instead they form giant lattices in which many ions arrange themselves into a regular structure.

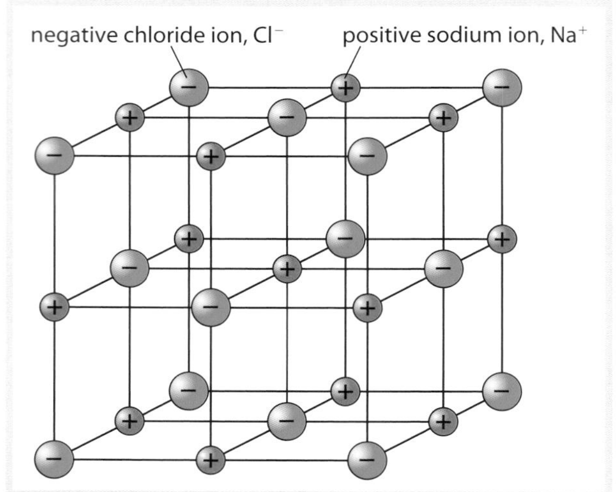

The arrangement of ions in a sodium chloride lattice.

The table below shows how the properties of ionic substances can be explained by their structure and bonding.

Property	Explanation
High melting point	The + and – ions attract each other strongly. A lot of energy is needed to separate them.
Conduct electricity when they are dissolved in water or melted	The ionic solid separates into ions which are free to move.
Often dissolve in water	Water is attracted to the charged ions and bonds to them.

Because they dissolve in water and conduct an electric **current**, ionic substances are used in electrical cells (or batteries). Batteries labelled 'alkaline' contain a solution of the ionic substance sodium hydroxide. This conducts the electrical current generated by a chemical reaction in the battery.

Simple covalent substances

strong covalent bond

weak force between molecules

Gaseous chlorine showing the strong covalent bonds within each molecule and weak forces between the molecules.

Although the bonding between the atoms in a covalent molecule is very strong, there are only very weak forces between the molecules. The bonding of simple covalent substances can be used to explain their properties.

Property	Explanation
Low melting and boiling point so they are often gases or liquids	The forces between molecules are very weak so not much energy is needed to separate them.
Do not conduct electricity even when melted	There are no charged ions in the substance to carry the current.
Not usually soluble in water, but they may dissolve in organic solvents like hexane	The molecules are not charged so water molecules are not attracted to them.

Giant covalent substances

Sand (silicon dioxide, SiO_2) contains silicon atoms and oxygen atoms bonded covalently. Silicon dioxide melts at around 1700 °C. This is higher than for many ionic substances. How do we explain this?

The way that silicon and oxygen bond means that they do not form a simple molecule. Each oxygen atom bonds to two silicon atoms and each silicon atom bonds to four oxygen atoms. They form a giant covalent structure.

All the bonds in the structure are strong covalent bonds. There are no weak forces in the structure so a lot of energy is needed to break the structure apart. This means that silicon dioxide is a very hard substance. Because it is such a hard substance and doesn't dissolve in water, it is often used in toothpaste as an abrasive to remove plaque from teeth. Other giant covalent structures include diamond and graphite.

= silicon atom

= oxygen atom

Structure of silicon dioxide – a giant covalent structure.

Activity A

1 Use the diagram showing the structure of diamond to explain some of its properties.
2 Diamond is used in industrial drills. Explain why it can be used in this way. Make sure that you use ideas about the structure and bonding in your explanation.

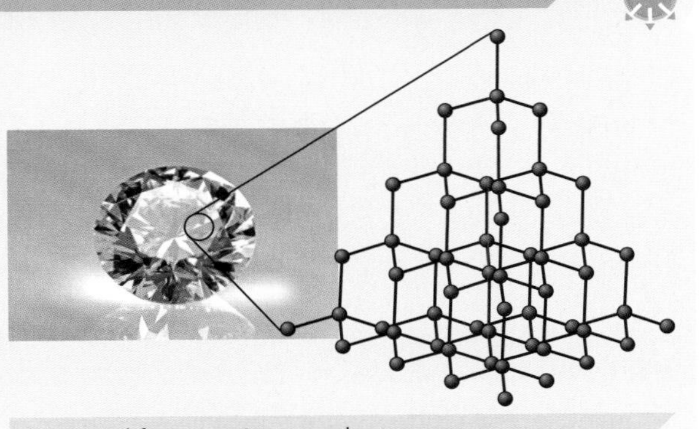

Diamond forms a giant covalent structure.

Assessment activity 2.1

| 2A.P2 | 2A.P3 | 2A.M2 | 2A.M3 | 2A.D2

You work in a pharmaceutical laboratory and you find that the labels on three bottles of chemicals in the laboratory have come off. All contain solid crystals. You know that the three substances are iodine, silicon dioxide and magnesium chloride, but you do not know which is which. One of these is ionic, one is simple molecular and one is giant molecular.

1 Prepare a briefing sheet that a new technician could use to decide which substance is which.
2 To make sure your technician understands the different types of bonding in the three substances, include diagrams of the structures and bonding.
3 Finally, describe some of the uses of these chemicals and how they can be related to the structure and bonding.

Tips

For 2A.P2 you should describe the differences in the properties of ionic and covalent substances.

For 2A.P3 you should draw some diagrams of the structure and electron arrangements in a range of covalent and ionic substances.

For 2A.M2 you will need to explain why these three types of substance have different properties.

For 2A.M3 you should also use the electron arrangement diagrams to describe how ionic and covalent substances form.

For 2A.D2 you will need to research some of the uses of these three substances and choose at least two uses that can be explained by using ideas about structure and bonding.

Just checking

1 List three properties of ionic substances.
2 Carbon dioxide has a simple covalent structure and silicon dioxide has a giant covalent structure. State one property that will be different for these two substances.

Lesson outcome

You should be able to describe and explain the properties of simple covalent, giant covalent and ionic substances.

2.3 Group 1 elements – properties and trends

Key terms

Physical property – Something we can observe about a substance that doesn't involve a chemical change; for example, its appearance or melting point.

Chemical property – Something we can observe about a substance that involves a chemical change; for example, how it reacts with water.

Trend – A change that occurs in a particular way, usually an increase or a decrease.

Take it further

In the table on the right, electrical conductivity is measured in milli-Siemens per metre (MS/m). The Siemens (S) is the unit of conductance and is equal to one ampere per volt. The symbol 'S' should not be confused with 's' for seconds.

Physical properties

Elements in the same group in the periodic table often have many **physical properties** in common.

Group 1 elements, the alkali metals:

- are soft, silver-grey metals
- have low melting points
- have high boiling points
- conduct electricity well.

But the properties aren't exactly the same for every element. Often there is a pattern or **trend** that you can see by looking at data for a property. The table below shows the physical properties of group 1 elements. As you can see from the table, there is a trend where the melting point decreases as you go down the group.

Element	Melting point (°C)	Boiling point (°C)	Electrical conductivity (MS/m)
Lithium	181	1342	11
Sodium	98	883	20
Potassium	64	760	16
Rubidium	39	688	48
Caesium	29	671	5

Activity A

1. Describe the trend in boiling point of group 1 elements.
2. Look at the data for electrical conductivity. Is there a simple trend in these data as you go down the group?
3. Use graph-plotting software to plot bar charts to show the trends in the data given in the table.

Chemical properties

Group 1 elements are very reactive. They will react quickly with gases like oxygen and chlorine to form ionic compounds.

Group 1 elements will also react with water. Potassium reacts with water to form hydrogen gas and an alkali, potassium hydroxide. The reaction produces heat which ignites the hydrogen gas.

All of the group 1 elements react to form the same kind of products but there is a trend in how fast and violent the reactions are. If a metal reacts very fast, chemists say it is very reactive.

Potassium reacting with water.

The trend is that the reactivity of group 1 elements increases as you go down the group in the periodic table.

Element	Observations when added to water
Lithium	Fizzing
Sodium	Rapid fizzing. The metal melts and floats on the water. Sparks are sometimes seen.
Potassium	The metal melts and may explode. The gas burns with a lilac flame.
Rubidium	Metal explodes and the gas bursts into flames.

Did you know?

No-one has ever seen a piece of francium, the element at the bottom of group 1 in the periodic table. In the whole of the Earth's crust there is less than 30 g at any one time.

Activity B 

1 Use the trend in reactivity and the periodic table to predict what would happen when caesium is added to water.
2 Use the Internet to find a video clip of this happening – was your prediction correct?

Link

You were introduced to electronic configuration in lesson 1.13.

Explaining the properties and trends

We can use ideas about the structure of the atoms in group 1 to explain their properties and trends.

As you can see from the diagrams below, there is one electron in the outer shell of each of the atoms in group 1. Group 1 elements are very reactive because the outer electron is easily lost to form an ion with a full outer shell.

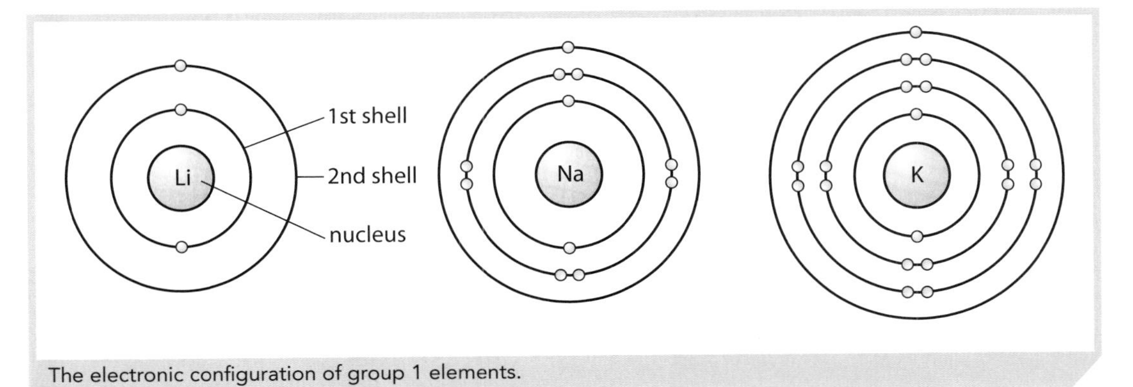

The electronic configuration of group 1 elements.

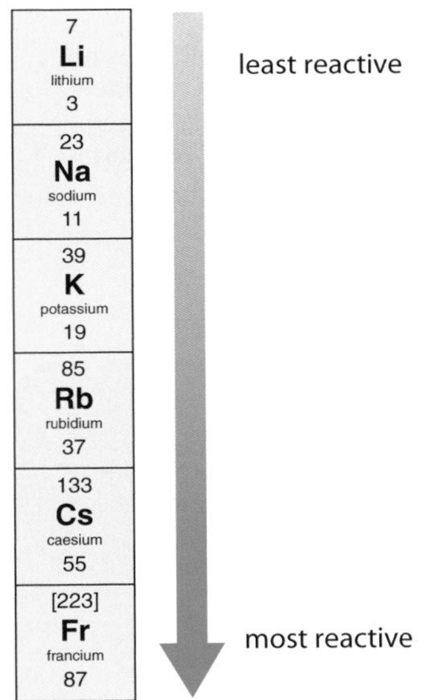

Trend in reactivity of group 1 elements.

The reactivity increases as you go down the group because the outer negative electrons are further away from the positive nucleus. It is easier to lose the electron from a larger atom because it is not as strongly attracted to the nucleus.

Just checking 

1 Which group 1 element has the highest melting point?
2 The reactivity of group 1 elements increases as you go down the group. Use ideas about electronic configuration to explain why.

Lesson outcomes

You should know about:

- the reactivity of group 1 elements with water
- the trends in the physical and chemical properties of group 1 elements in relation to their electronic configuration.

Key term

Displacement – A reaction in which a more reactive element displaces a less reactive element from a solution of its compound.

Activity A

Use graph-plotting software to draw bar charts to show the trends in melting point and boiling point as you go down group 7.

Safety and hazards

The vapours of the halogens are very toxic. Liquid bromine is corrosive and solid iodine is harmful. However, in low concentrations, solutions of some of the halogens, chlorine and iodine, can be used to sterilise water.

Link

You will find out more about word equations and chemical equations in lesson 2.7.

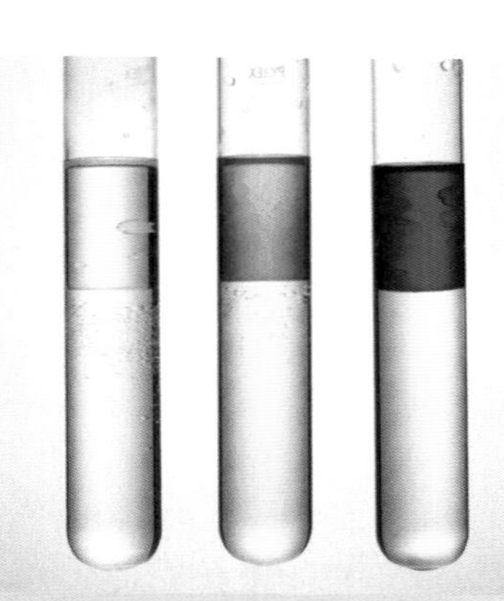

The top layer shows the colour of different halogens when dissolved in hexane. From left to right: chlorine, bromine, iodine.

Physical properties

Group 7 elements are non-metals made up of molecules such as fluorine, (F_2), chlorine (Cl_2), bromine (Br_2) and iodine (I_2). These elements are called **halogens** and they have several physical properties in common. All halogens:

- have quite low melting and boiling points
- are coloured substances when pure or in solution
- dissolve well in organic solvents like hexane, but not very well in water.

The table below shows the physical properties of some group 7 elements.

Element	Appearance at room temperature	Appearance when dissolved in hexane	Melting point (°C)	Boiling point (°C)
Chlorine	Pale-green gas	Pale-green solution	−101	−34
Bromine	Red liquid	Orange solution	−7	59
Iodine	Black solid	Purple solution	114	184

Chemical properties

Group 7 elements show a trend in reactivity like group 1 elements. Instead of reacting with water, like group 1 elements, halogens react with a compound containing a different group 7 element. A **displacement** reaction happens:

chlorine	+	potassium bromide	→	bromine	+	potassium chloride
Cl_2	+	2KBr	→	Br_2	+	2KCl

Chlorine is more reactive than bromine so it displaces the bromine from the potassium bromide compound. When halogens react to form compounds, these new compounds are often called **halides**. The table below summarises what reactions take place.

Substance added	Observation when added to potassium chloride	Observation when added to potassium bromide	Observation when added to potassium iodide
Chlorine (pale green)	–	Orange (bromine formed)	Purple (iodine formed)
Bromine (orange)	Stays orange	–	Purple (iodine formed)
Iodine (purple)	Stays purple	Stays purple	–

The results in the table show you that the reactivity of the group 7 elements decreases as you go down the group.

Explaining the properties and trends

The electronic structure of group 7 elements helps us to explain their properties.

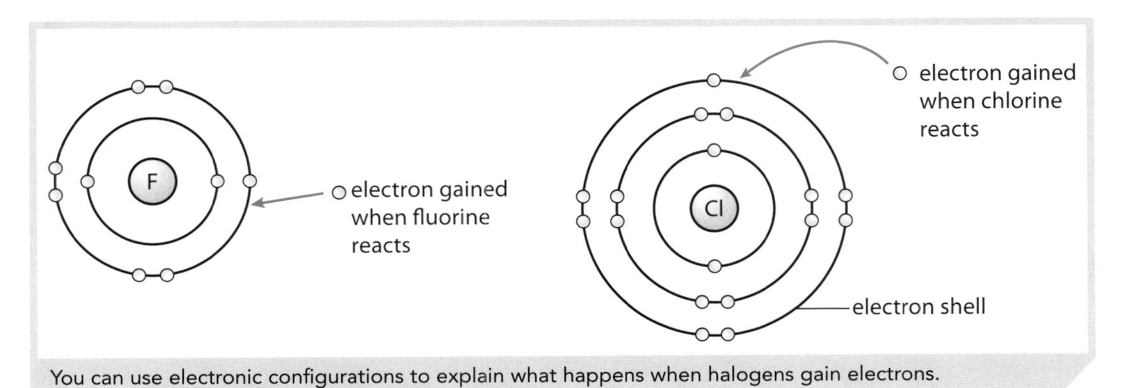

You can use electronic configurations to explain what happens when halogens gain electrons.

Group 7 elements are quite reactive because only one extra electron is required to form an ion with a full outer shell of electrons. However, the reactivity decreases as you go down the group because the atoms at the bottom of the group have more shells and the outer shell of electrons is further from the nucleus. It is harder to gain an electron into this shell because it is not so strongly attracted to the nucleus.

Remember

Chlorine and chloride mean different things. Chlorine is an element and is a reactive green gas. Chloride is the ion which forms when chlorine reacts. It is very stable because it has a full outer shell of electrons.

Assessment activity 2.2

| 2A.P1 | 2A.M1 | 2A.D1

Fluorine and caesium are two very reactive elements. They are not found in school laboratories although they are used in the chemical industry.

Prepare a leaflet about these elements aimed at new laboratory technicians who may not be familiar with these substances. You should compare fluorine and caesium to other elements in the same groups. Then you need to explain how trends in the physical and chemical properties of group 1 and group 7 elements help chemists to predict the properties of these two elements.

Tips

For 2A.P1 you should be able to give examples of at least two other elements from group 1 and two other elements from group 7 and describe their physical and chemical properties.

For 2A.M1 you need to describe the trends in at least one physical and one chemical property to help you to predict the properties of caesium and fluorine.

For 2A.D1 you will need to be able to explain these trends using ideas about electronic structure.

Link

Lesson 2.3 discusses the properties and trends of group 1 elements.

Just checking

1. Sarah adds a solution of chlorine to a solution of potassium iodide. What will she observe? What kind of reaction has happened?
2. Describe the trend in reactivity of group 7 elements as you go down the group.

Lesson outcomes

You should:

- know about the trends in physical and chemical properties of group 7 elements, and their displacement reactions
- be able to relate the trends in chemical properties of group 7 elements to their electronic configuration.

2.5 Physical properties and uses of chemicals

Get started

Here are some properties of the substance hexane. Decide whether they are physical properties or chemical properties. Hexane:

- boils at 69 °C
- burns in air to form carbon dioxide and water
- is a colourless liquid.

Melting and boiling point

You can look up or measure the melting point and boiling point of a substance to help you decide whether it is a solid, liquid or gas at room temperature.

For example, ethanol is used as a fuel. It has a melting point of −114 °C and a boiling point of 78 °C. Room temperature is about 20 °C. So at room temperature there is enough energy to melt ethanol but not to boil it. This means that ethanol is a liquid at room temperature. Below −114 °C ethanol is a solid.

Electrical conductivity

Some substances can conduct an electric current when a voltage is applied to them. You can look up the electrical conductivity of a substance or investigate it practically.

Metals, especially copper and aluminium, have very high electrical conductivity – they are called conductors. If a substance does not conduct electricity well it is called an insulator.

Silicon is used in the electronics industry to make silicon chips like those in your computer.

Semiconductors

Silicon is an element in group 4 of the periodic table. This means that it is half way between being a metal (like those in group 1) and a non-metal (like those in group 7). So it is a **semiconductor** – it has moderate electrical conductivity.

High-voltage power lines are made of metal to allow a current to flow easily without losing energy. The lines are attached to the pylons by discs of ceramic insulators.

Activity A

Look at the data in this table which show the physical properties and costs of some metals.

Metal	Electrical conductivity (MS/m)	Strength (relative to Al = 100)	Density (g/cm^3)	Cost (2012) (£ per tonne)
Copper	60	300	8.9	6000
Aluminium	35	100	2.7	2200
Iron	10	1000	7.9	520

Copper, aluminium and iron can be used in overhead power cables.

1 Old power cables were made from copper. Explain why copper is a good choice for a power cable.

2 Most modern power cables are now made from a mixture of iron and aluminium. Explain why this combination of metals is more suitable.

Thermal conductivity

Take-away hot drinks are sometimes served in expanded polystyrene cups. The polystyrene is a thermal insulator – it has low thermal conductivity so heat cannot flow through it.

Metals have high thermal conductivities but many other substances, like polymers, ceramics and all gases, have low thermal conductivities.

Did you know?

Water is often used to cool chemical reactors by transferring heat away from the reactor.

However, the nuclear reactors in modern nuclear power stations produce so much heat that liquid sodium is used. Sodium has a much higher thermal conductivity than water.

Solubility

When you studied group 7 elements you found that they dissolved well in organic solvents like hexane, but they didn't dissolve very well in water. We say that these elements have a high **solubility** in hexane but a low solubility in water.

Viscosity

Many liquids, like water, flow well when you pour them. Other liquids, like olive oil or motor oil, seem much 'thicker' and don't flow so well. These liquids are called viscous and you can measure their **viscosity**. Water has low viscosity, while olive oil has high viscosity.

This lubricating oil is quite viscous, so it doesn't pour very well.

Assessment activity 2.3

2B.P4 | 2B.M4 (part) | 2B.D3 (part)

Motor oil is a mixture of several different liquids used to lubricate the moving parts of an engine. It also helps to clean the engine and cool it.

1 Use research from websites or the labels on containers of oil to find the physical properties of a sample of motor oil or of the liquids which are mixed to produce it.
2 Produce a poster that explains these properties and why they are important in the use of motor oil.

Tips

For 2B.P4 you will need to list the important physical properties of motor oil and then describe and explain how these physical properties make motor oil suitable for use in a car engine. By discussing and explaining the link between the properties and uses you will also provide part of the evidence needed for 2B.M4 and 2B.D3.

Lesson outcomes

You should:

- understand how the uses of chemicals are based on their physical properties (electrical and thermal conductivity, melting and boiling points, solubility, viscosity)
- know about the use of silicon in computer-chip technology.

Just checking

1 Butanol has a melting point of −90 °C and a boiling point of 118 °C. Will it be a solid, liquid or gas at room temperature (20 °C)?
2 Name one type of substance which has high electrical conductivity.
3 Many saucepans are made of metal but have a ceramic handle. Explain why.

2.6 Chemical properties and uses of chemicals

Get started

In lesson 1.20 you discovered how indigestion remedies are used to treat stomach ache caused by too much acid in the stomach. Calcium carbonate is used in many indigestion remedies.

Discuss why calcium carbonate is a suitable choice for an ingredient in indigestion remedies. You will need to think about its chemical properties.

Sodium azide in car airbags

Nowadays cars are fitted with airbags which protect you if the car is involved in a collision. Each airbag contains about 100 g of a chemical called sodium azide. There is also a small amount of a detonator compound. If there is an accident, the shock of the collision causes sensors in the car to send an electrical signal to the detonator. This reacts and releases heat which causes the sodium azide to break down.

Airbags cushion passengers from impact in an accident.

One hundred grams of sodium azide, which is enough to fill a small cup, produces over 50 litres of nitrogen gas. It does this incredibly quickly – the airbag inflates in less than 50 milliseconds.

Sodium azide is a white ionic solid which has the chemical formula NaN_3. It looks just like ordinary table salt (sodium chloride) and normally it is quite stable and unreactive at room temperature. But when it is heated to 300 °C it breaks down in a reaction called **thermal decomposition**.

$$\text{sodium azide} \rightarrow \text{sodium} + \text{nitrogen}$$
$$2NaN_3 \rightarrow 2Na + 3N_2$$

Safety and hazards

Sodium is very reactive and reacts with water to form sodium hydroxide, which is very corrosive. So airbag manufacturers mix the sodium azide with other chemicals to react with the sodium when it forms.

Argon for arc welding

Argon is an element from group 0 of the periodic table. The elements in this group are called the **noble gases** and they are very unreactive. This means argon will not react with oxygen or metals, even at very high temperatures.

Argon is used in arc welding. In this process, pieces of metal are joined together by melting them with an electric current and letting them cool. If hot metal is in contact with air it can react. This can be dangerous or affect the weld joint. To prevent this, a stream of argon is passed over the joint during welding.

Activity A

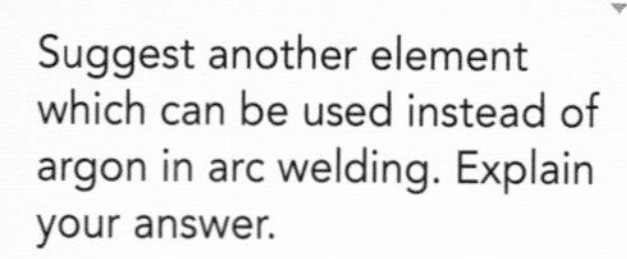

Suggest another element which can be used instead of argon in arc welding. Explain your answer.

Carbon dioxide in fire extinguishers

Carbon dioxide is a gas made up of covalent molecules with the formula CO_2. Because the covalent bonds in each molecule are strong, carbon dioxide is quite unreactive, except at very high temperatures.

Carbon dioxide is used in fire extinguishers. Fires happen when materials like wood, paper or plastics react with the oxygen in the air at high temperatures. These substances do not react with carbon dioxide. So, when carbon dioxide is sprayed from a fire extinguisher over the burning material it covers the material, stopping oxygen reaching it. The fire will then quickly go out.

A CO_2 fire extinguisher in action.

Did you know?

Carbon dioxide is denser than air, so when it is sprayed on a fire it will sink, pushing the air out of the way.

Assessment activity 2.4

2B.P4 | 2B.P5 | 2B.M4 (part) | 2B.D3 (part)

You are a chemist working for a company which manufactures rocket fuel for rockets that launch communication satellites into space. You are considering developing new rocket fuels. There are three possible fuels to consider: hydrogen, hydrazine and gunpowder (a mixture of potassium chlorate, sulfur and carbon). Note that rockets carry a supply of oxygen to react with the fuel.

Research what properties are needed for a rocket fuel. Write a report on the properties of the three possible fuels, explaining which one will be the best choice.

Grading tips

For 2B.P4 and 2B.P5 you will need to describe some physical properties and also some chemical properties of the three substances and explain how these properties are important when they are used as a fuel.

By explaining and discussing the link between the properties and uses, you will also provide part of the evidence needed for 2B.M4 and 2B.D3.

For 2B.D3 you will need to choose one fuel and explain why the properties of the fuel make it the most appropriate choice.

Just checking

1 Look at the following descriptions of the properties of substances. Decide whether each property described is a physical or a chemical property.
  (a) Sodium reacts with water to produce hydrogen gas and sodium hydroxide.
  (b) Glycerol is a very viscous substance.
  (c) Ethanol dissolves in water.
  (d) Nitrogen is an unreactive gas.

Lesson outcome

You should understand how the uses of chemicals are based on their chemical properties.

2.7 Word equations and chemical equations

Get started

Chemical reactions are happening all around you in your daily life. Discuss with other learners chemical reactions which you have seen happening today (outside your school or college laboratory).

Key terms

Reactants – The substances that react together in a chemical reaction.

Product(s) – The new substance(s) formed by a chemical reaction.

State symbol – Abbreviation used to show the physical state of each substance in a reaction:
- (s), solid
- (l), liquid
- (g), gas
- (aq), aqueous.

Link

The ideas about equations in this section link with lesson 1.17.

Remember

Word equations always include the full chemical names of the reactants and products. So although sodium chloride is often called 'salt' we always use its chemical name.

Chemical reactions

You have probably seen quite a lot of chemical changes happening in your school or college laboratory. But many chemical changes happen inside living organisms or in everyday situations like cooking.

You can recognise chemical changes because new substances are formed, with different properties. So you could see a colour change, a gas being given off or heat being given off when a fuel burns.

Word equations

When sodium and chlorine react together a new compound, sodium chloride, is formed. Sodium is in group 1 of the periodic table and chlorine is in group 7 of the periodic table. Both are reactive elements, so this reaction is quite violent. The reaction can be shown as a word equation. The arrow shows you that a chemical change happens:

sodium + chlorine → sodium chloride
reactants **product**

Be particularly careful about spellings: chlorine, chloride and chlorate all mean different things.

This is sodium reacting with chlorine to make sodium chloride (salt).

Activity A

Look at the descriptions of chemical reactions below. Use the descriptions to write word equations for the reactions.
- When sodium is added to water it gives off hydrogen and a solution of sodium hydroxide is left behind.
- Chlorine displaces bromine from a solution of potassium bromide. Potassium chloride is also formed.

Balanced chemical equations

Chemists often write balanced chemical equations. These give more information about the reaction and:
- the formulae of the reactants and products
- the number of particles of reactants and products involved in the reaction
- the physical state of each substance, solid, liquid, gas or dissolved in water, shown by the **state symbols**: (s), (l), (g) and (aq).

In a balanced chemical equation, the reactants must have the same number of atoms of each type as the products. Here is a balanced chemical equation for the combustion of hydrogen:

$$2H_2\,(g) + O_2\,(g) \longrightarrow 2H_2O\,(l)$$

hydrogen + oxygen ⟶ water

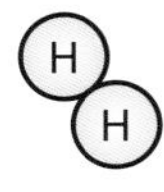

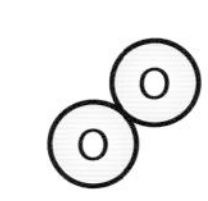

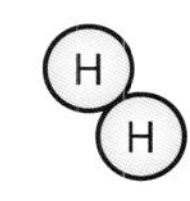

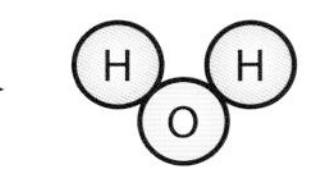

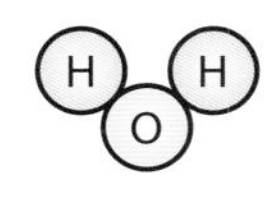

Drawing the molecules helps you to count the number of atoms of each type involved in the reaction.

Worked example

The balanced chemical equation for the reaction between magnesium and hydrochloric acid is:

$$Mg\,(s) + 2HCl\,(aq) \rightarrow MgCl_2\,(aq) + H_2\,(g)$$

What does this tell you?

- There are 1 Mg atom, 2 H atoms and 2 Cl atoms involved in the reaction.
- The reactants are solid magnesium and hydrochloric acid dissolved in water.
- The products are magnesium chloride dissolved in water and hydrogen gas.

Reversible and irreversible chemical changes

Nitrogen dioxide (NO_2) is a brown gas and is one of the pollutants in car exhausts. Two nitrogen dioxide molecules can join to make a new product called dinitrogen tetroxide (N_2O_4), a colourless gas. If N_2O_4 is heated, it splits back up into NO_2 molecules. So we say that the reaction is **reversible**.

$2NO_2$		N_2O_4
nitrogen dioxide		dinitrogen tetroxide
brown		colourless

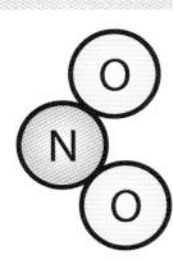

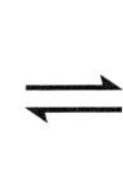
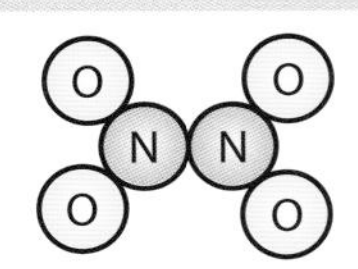

The ⇌ symbol shows you that the reaction is reversible. Most reactions aren't easily reversible which is why the forward arrow (⟶) is used in most chemical equations.

Just checking

1 Look at this balanced chemical equation:

$$Cl_2\,(g) + 2NaBr\,(aq) \rightarrow Br_2\,(aq) + 2NaCl\,(aq)$$

(a) Name the products of the reaction.

(b) How many atoms of sodium are involved in this reaction?

(c) Is the reaction reversible? Explain how you can tell.

(d) In what state is the reactant sodium bromide?

Activity B

Ethanol, a useful fuel, can be made from ethene and water under certain carefully controlled conditions. The balanced chemical equation is:

$$C_2H_4\,(g) + H_2O\,(g) \rightleftharpoons C_2H_5OH\,(g)$$

Write down everything that you can work out about the reaction from this equation.

Lesson outcomes

You should:

- be able to write word equations and simple balanced equations (including state symbols)
- recognise reactants and products in a reaction
- know the difference between reversible and irreversible changes.

2.8 Reactions and reaction rates

Get started

You can sometimes tell that chemical reactions are happening because you see changes happening. List some of the observations that you might make to tell you that a chemical reaction is happening.

Link

This links with your work on neutralisation in lessons 1.15 to 1.17.

Activity A

For each of the chemical equations shown:

- name the products and reactants
- write down the number of atoms of each type involved in the reaction.

Types of reaction

You have come across several different types of reaction in the previous sections.

- **Displacement**: A more reactive element displaces a less reactive element from a solution of its compound.

$$Cl_2\,(aq) + 2KI\,(aq) \rightarrow I_2\,(aq) + 2KCl\,(aq)$$

- **Combustion**: A substance burns in oxygen to form oxides.

$$2H_2\,(g) + O_2\,(g) \rightarrow 2H_2O\,(l)$$

Hydrocarbons, which contain carbon and hydrogen atoms, burn to form carbon dioxide and water:

$$CH_4\,(g) + 2O_2\,(g) \rightarrow CO_2\,(g) + 2H_2O\,(l)$$

- **Neutralisation**: An acid is neutralised by an alkali, a metal or a metal carbonate.

$$2HCl\,(aq) + Mg\,(s) \rightarrow MgCl_2\,(aq) + H_2\,(g)$$

$$2HCl\,(aq) + CaCO_3\,(s) \rightarrow CaCl_2\,(aq) + CO_2\,(g) + H_2O\,(l)$$

Reaction rates

Some of the reactions you have studied happen very fast. For example, hydrogen and oxygen react together in a rapid explosion like the one shown on the opening page of this unit.

Iron rusting is an example of a chemical reaction.

Other reactions happen more slowly. For example, when you add magnesium ribbon to hydrochloric acid it can take several minutes for the magnesium to react completely. Reactions like rusting (iron reacting with oxygen and water) happen even more slowly – over a period of months or years.

We talk about substances having a fast or slow rate of reaction.

Investigating reaction rates

You can investigate the rate of a chemical reaction by measuring how fast a product is formed. This is easy to do if the product is a gas. Reactions of acids with metals, or acids with carbonates, produce gases.

When magnesium ribbon is added to hydrochloric acid it produces hydrogen gas which can be used as a measure of the rate of the reaction.

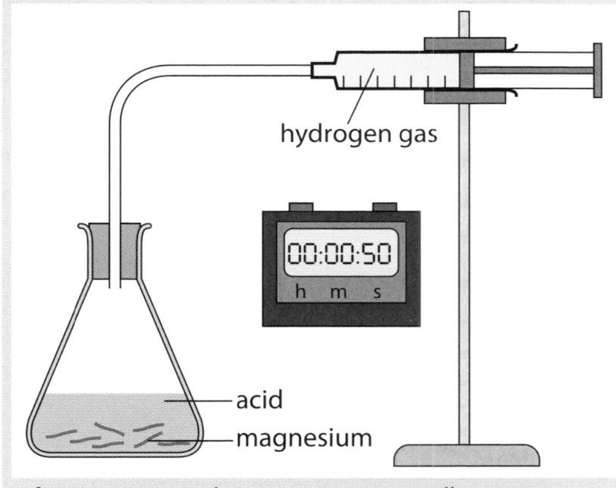

If a reaction produces gas you can collect it in a gas syringe.

magnesium + hydrochloric acid → magnesium chloride + hydrogen

$$Mg + 2HCl \rightarrow MgCl_2 + H_2$$

Worked example

A reaction produces 50 cm³ of gas in 20 s. What is the rate of the reaction?

$$\text{Rate of reaction} = \frac{\text{volume of gas produced}}{\text{time}} = \frac{50\,cm^3}{20\,s} = 2.5\,cm^3/s$$

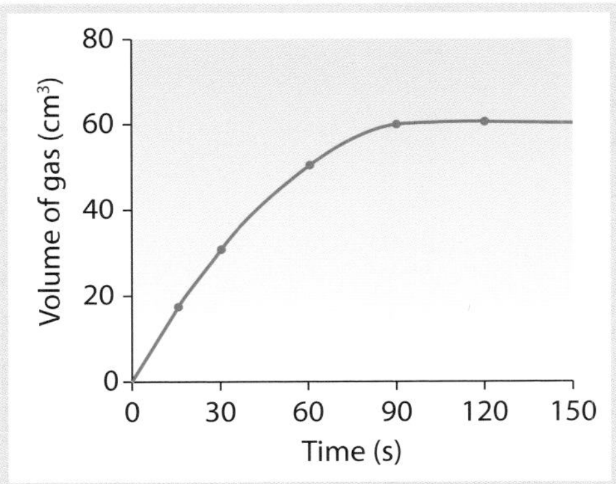

You can use graphs like this to find the rate of a chemical reaction.

If you measure the volume of a gas produced every 10 seconds you can plot a graph to show what happens during the reaction. A good way of analysing the graph is to find out how long it takes to produce a certain volume of gas.

Activity B

Use the graph shown to find the time taken to produce 40 cm³ of gas. What is the rate of the reaction?

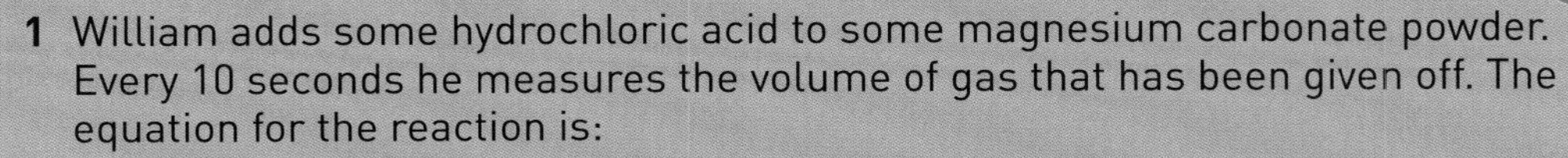

Just checking

1 William adds some hydrochloric acid to some magnesium carbonate powder. Every 10 seconds he measures the volume of gas that has been given off. The equation for the reaction is:

$$2HCl\,(aq) + MgCO_3\,(s) \rightarrow MgCl_2\,(aq) + CO_2\,(g) + H_2O\,(l)$$

(a) What is the gas given off?

(b) How could William collect the gas to measure its volume?

(c) William collects 60 cm³ of gas in 40 seconds. What is the rate of the reaction (in cm³/s)?

Lesson outcomes

You should be able to recognise reactants and products in a reaction (displacement, combustion, neutralisation), and use reaction rate graphs.

2.9 Rates of reaction – concentration, pressure and surface area

Get started

Imagine you are going to drop a sugar lump into a mug of tea. Make a list of the things you could do to make the sugar lump dissolve as fast as possible.

Key term

Concentration (of solution) – The amount of a solute dissolved in a certain volume of solution. It can be measured in the units g/dm^3 (grams per cubic decimetre).

Concentration

If a solution is concentrated it means that there is a lot of a substance dissolved in a small volume of solvent. Reactions involving solutions usually happen faster if the solution is more concentrated.

You could do an experiment to measure the time taken for a piece of magnesium ribbon to dissolve in different **concentrations** of hydrochloric acid.

Concentration of hydrochloric acid (g/dm^3)	Time taken in seconds for a 2 cm piece of magnesium ribbon to dissolve
40	21
30	30
20	43
10	88

A short time means a fast rate, so the results in the table show that at higher concentrations of hydrochloric acid the reaction proceeds at a faster rate.

Activity A

Plot a graph of the results in the table. Time should go on the vertical axis and concentration on the horizontal axis.

Use the graph to predict how long a 2 cm piece of magnesium would take to dissolve in a solution of hydrochloric acid with a concentration of $36.5\ g/dm^3$.

Pressure

If a reaction involves gases rather than solutions then it will happen faster if the gas is at high pressure. This is because at high pressure there are a lot of gas molecules in a small volume.

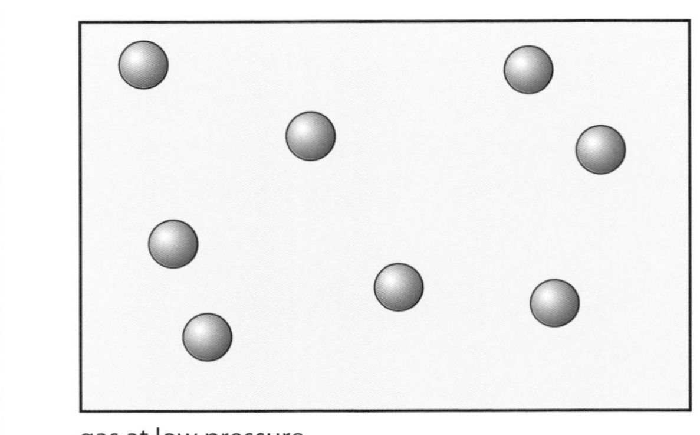

gas at low pressure

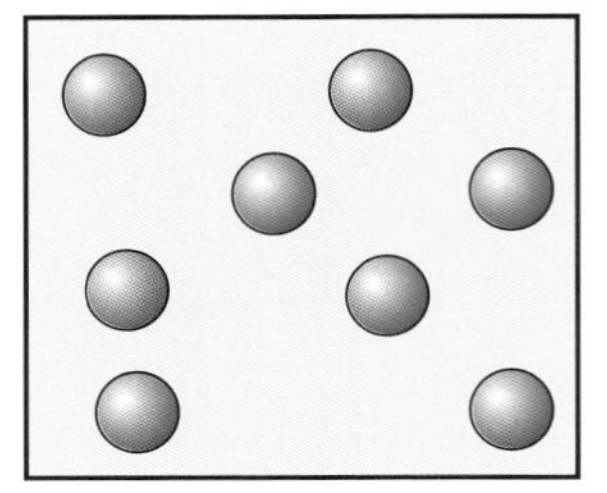

gas at high pressure

The effect of pressure on gas molecules.

Surface area and particle size

If a solid is ground up into a fine powder, its surface area increases. Very fine powders can react very quickly because they have a large surface area.

The graph shows the results of an experiment that measured the volume of gas given off by marble chips of different sizes when they react with hydrochloric acid.

The reaction with the small chips produces the same amount of gas in a shorter time, so the rate of the reaction is faster.

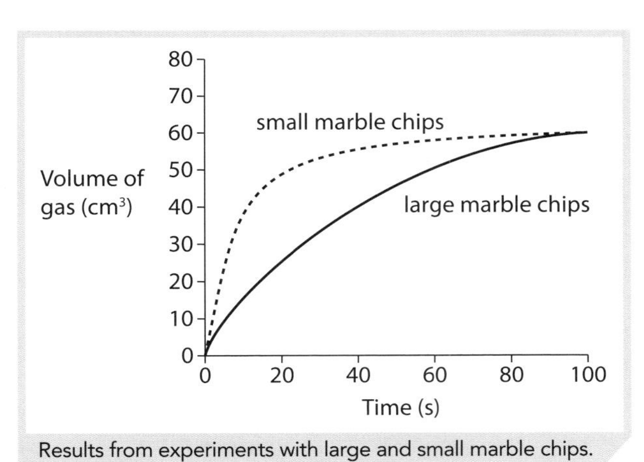

Results from experiments with large and small marble chips.

Case study

Emily works for a company that manufactures descalers to use in household appliances.

'Many people live in hard water areas. This means that the water contains ions like calcium and hydrogen carbonate. If the water is heated up in a kettle, boiler or coffee-maker, the ions react together to form a white solid known as limescale (calcium carbonate). Limescale forms a solid layer which can cover the heating element or block up pipes and tubes. Our descalers will react with the limescale and remove it.

Limescale on a kettle element.

Descalers contain solutions of acids, like citric acid or ethanoic acid. We are trying to develop descalers which work as fast as possible but are safe to use in the house.'

1. The reaction of limescale with descalers in these household devices is much slower than when descaler is added to powdered calcium carbonate in the laboratory. Explain why.
2. What changes might Emily's company make to the descaler solution to try and speed up the reaction?
3. Suggest a reason why they may choose not to make these changes.

Safety and hazards

Factories that make products like custard powder and powdered milk must take special precautions to avoid any dust escaping into parts of the factory where there may be sparks.

Custard powder can catch fire and, because the surface area of the fine powder is so large, it would create an explosion as it burned.

Activity B

There are several ways in which you can use the graph on this page to conclude that small marble chips react faster than large marble chips.

1. Which line is steeper at the start of the reaction?
2. Calculate the time taken for 40 cm^3 of gas to be produced for each size of marble chips.
3. Estimate the time taken for each reaction to be over.

Take it further

Chemists often analyse graphs like this by working out the gradient of the graph early on in the reaction.

Use the graph to find its gradient in the first 10 s for each size of chip. The gradient is equal to the rate of the reaction.

Lesson outcome

You should understand the effect of surface area and concentration on the rate of chemical reactions.

2.10 Rates of reaction – temperature and catalysts

Key term

Catalyst – A substance that speeds up a chemical reaction but is not used up, so it can be used again.

Did you know?

For many reactions, a 10 °C increase in temperature causes the rate of reaction to roughly double.

Activity A

Use the graph to find the rate of reaction at **(a)** 15 °C and **(b)** 35 °C. What do you notice?

Temperature

You have often heated reactions to speed them up. Outside the laboratory you put food in the fridge because the low temperature slows down the reactions which cause decay. You could do an experiment in which you measured the rate of a reaction at different temperatures. A good way of presenting the results would be to plot a graph of rate against temperature as shown below. You can see that changing the temperature by a few degrees can have a noticeable effect on the rate of the reaction.

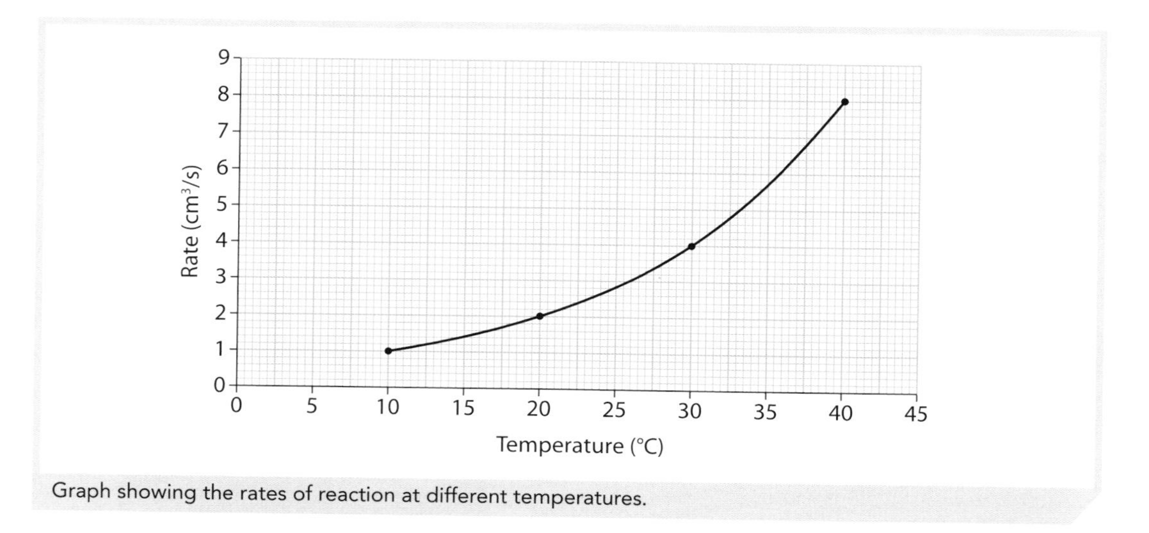

Graph showing the rates of reaction at different temperatures.

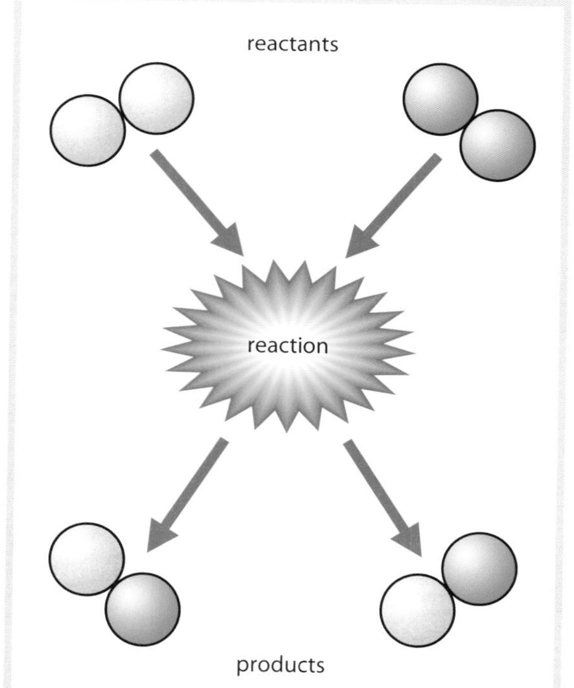

Two molecules collide and react if there is enough energy to allow the atoms to rearrange into products.

Catalysts

Hydrogen peroxide has the chemical formula H_2O_2. It is like water with an extra oxygen atom. If you look at a solution of hydrogen peroxide you can sometimes see small bubbles appearing. The hydrogen peroxide is breaking down into simpler substances.

hydrogen peroxide → water + oxygen

$$2H_2O_2\,(aq) \rightarrow 2H_2O\,(l) + O_2\,(g)$$

If a small amount of manganese dioxide is added, the bubbling becomes very violent. The reaction rate has increased enormously. The manganese dioxide is a **catalyst**. It is not used up in the reaction so it can be filtered off and used again.

Collision theory

For a reaction to happen between two substances, the particles of the substances must collide with enough force (or energy) for them to break up and re-bond to form new products. This is called the **particle collision theory**.

You can use collision theory to help you understand why different factors affect the rates of reactions.

Factor	Effect
The concentration of a solution is increased.	Particles are closer together and collide more often.
The surface area of a solid is increased.	There are more particles available at the surface for other particles to collide with.
The temperature is increased.	The particles have more energy so when they collide they are more likely to react.
A catalyst is added.	The catalyst lowers the amount of energy needed for particles to react so more of the collisions result in a reaction.

Assessment activity 2.5

| 2C.P6 | 2C.P7 | 2C.M5

You are a laboratory technician carrying out a survey of the chemical reactions used in a chemical factory. The three processes are:

- the manufacture of bromine: Cl_2 (g) + 2KBr (aq) → Br_2 (l) + 2KCl (aq)
- the hydration of ethene: C_2H_4 (g) + H_2O (g) → C_2H_5OH (g)
- the **Contact process**: $2SO_2$ (g) + O_2 (g) ⇌ $2SO_3$ (g)

1 For each process, use the equation to write down:
   (a) the names of the reactants and products
   (b) the states of the reactants and products
   (c) whether the reaction is reversible or not
   (d) the number of atoms of each type involved in the reaction.
2 Use research to find out whether a catalyst is used in the process. If it is, find out its name or chemical formula.
3 For each reaction, describe the factors that will affect the rate of the reaction and explain why they affect the rate.

Tips

For 2C.P7 you will simply need to follow the instructions in question 1 very closely. You may choose to present your answers in the form of a table.

For 2C.P6 you need to describe which factors affect the rate of each reaction. Be careful only to include relevant factors – if a reaction doesn't use a catalyst then you shouldn't mention it as a factor. You should decide how each factor affects the rate of the reaction.

Finally, for 2C.M5 you need to use collision theory to explain why each factor affects the rate of these industrial reactions.

Activity B

Jamie is investigating what happens when sodium carbonate powder reacts with hydrochloric acid solution. The reaction happens too quickly for him to collect all the carbon dioxide gas.

1 List three changes he could make to slow the reaction down.
2 For each change, use collision theory to explain why the reaction is slower.

Lesson outcomes

You should:

- understand the effect of catalysts and temperature on the rate of chemical reactions
- be able to use collision theory to explain why various factors affect the rate of reaction.

Key terms

Atom economy – A way of measuring the amount of atoms in the reactants that become useful products.

Yield – The mass of product made in a reaction.

% yield – A way of comparing the actual yield with the predicted yield for a reaction.

Feeding the world: fixing nitrogen

The world's population is increasing all the time. To try and produce the extra food needed to feed the world's population, farmers are relying more and more on nitrogen-based fertilisers to put nutrients back into the soil.

The first stage in the manufacture of nitrogen-based fertilisers is the reaction of nitrogen from the air with hydrogen from natural gas. Under special conditions nitrogen and hydrogen react to make ammonia in a reaction called the **Haber process**:

$$\text{nitrogen} + \text{hydrogen} \rightleftharpoons \text{ammonia}$$
$$N_2 + 3H_2 \rightleftharpoons 2NH_3$$

Ammonia is a very important chemical. It is used to make ammonium nitrate which is used as a fertiliser.

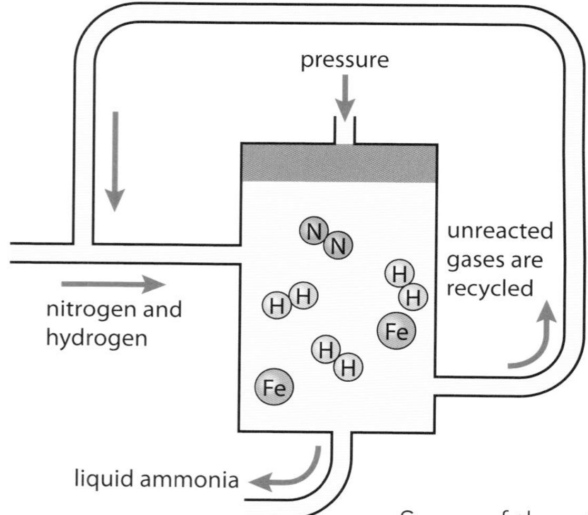

The Haber process.

Atom economy

The Haber process has a high **atom economy** because all the atoms from the reactants end up in the ammonia product rather than in other waste products.

$$\text{Atom economy} = \frac{\text{mass of atoms in the desired product}}{\text{total mass of atoms in all products}} \times 100$$

Because there are no waste products formed, the atom economy of the Haber process is 100%. Another way of producing ammonia is by breaking down urea:

$$\text{urea} + \text{water} \rightarrow \text{ammonia} + \text{carbon dioxide}$$
$$CO(NH_2)_2 + H_2O \rightarrow 2NH_3 + CO_2$$

Some of the atoms in the reactants end up in the waste product carbon dioxide. This means that the atom economy is lower. A low atom economy means that there are more waste products produced, which may be hazardous or expensive to dispose of.

Conditions for the Haber process

Industrial chemists try and choose reactions which have a high atom economy. Then they must choose conditions to try and make the process as cost-effective as possible. The table shows some of the factors they need to think about.

Factor	Reason
Rate	A fast rate means that a lot of product can be made in a short time.
Yield	A high yield means that smaller amounts of reactants can be used to make the same amount of product.
Fuel costs	Using high temperatures and pressures needs a lot of energy. Under these conditions, the fuel costs would be high.
Safety	Using high pressures can increase the risk of explosions or leaks. This means that expensive thick-walled reactors must be used.

Worked example

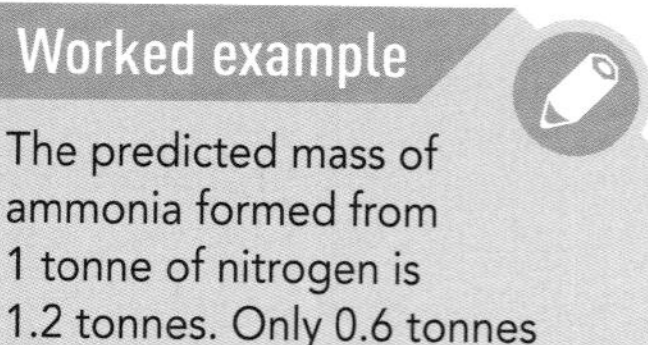

The predicted mass of ammonia formed from 1 tonne of nitrogen is 1.2 tonnes. Only 0.6 tonnes of ammonia is actually formed. What is the % yield?

$$\frac{0.6 \text{ tonnes}}{1.2 \text{ tonnes}} \times 100 = 50\%$$

The conditions that are used for the Haber process are:

- a moderately high temperature of 400 °C
- a high pressure of 200 atmospheres
- an iron catalyst.

Activity A

Temperatures above 400 °C and pressures above 200 atm are not usually used in the Haber process. Use ideas about fuel costs and safety to suggest some reasons why. What would happen to the rate if a catalyst wasn't used?

Rate and reaction yield

Increasing the pressure and temperature will make the reaction go much faster. However, temperature and pressure also affect the **yield**, as shown in the graph.

The yield of the reaction is the mass of product made in the reaction.

You can predict how much product should be made in a reaction (the predicted yield) and then calculate the **% yield** of the actual process.

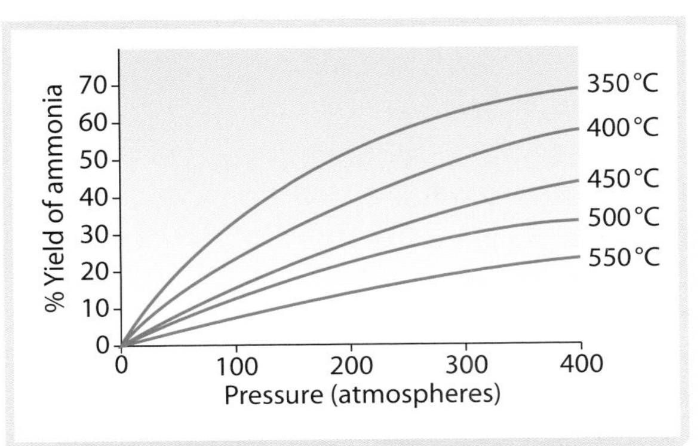

You can see from this graph how the yield changes when both pressure and temperature are changed.

$$\% \text{ yield} = \frac{\text{mass of product made (actual yield)}}{\text{predicted mass of product (predicted yield)}} \times 100$$

The % yield of the Haber process is often only about 30%. Possible reasons for low yields are:

- the reaction is reversible so some of the product is converted back to reactant
- some of the products have been lost during separation or purification.

Assessment activity 2.6

| 2C.M6 | 2C.D4

The Haber process and the Contact process are both important industrial reactions. The equations for these reactions are:

Haber process: $N_2\,(g) + 3H_2\,(g) \rightleftharpoons 2NH_3\,(g)$

Contact process: $2SO_2\,(g) + O_2\,(g) \rightleftharpoons 2SO_3\,(g)$

These processes both have an atom economy of 100%. The Haber process is usually operated with a yield of 30% whereas the yield of the contact process is about 90%.

You are a chemical engineer at an ammonia plant. You have been asked by the manager to explain the information above and to investigate whether the ammonia can be made more cheaply.

Write a report to explain to your manager the meanings of the terms 'atom economy' and 'yield'. Explain how the yields of these two reactions can be different, even though the atom economy is the same.

Your manager has suggested doing the reaction at 100 atm and 200 °C to make the ammonia more cheaply. Design a presentation to explain whether or not this is likely to be a good idea.

Tips

For 2C.M6 you will need to use the equation for atom economy to explain how these reactions have an atom economy of 100% – either with or without a calculation.

For 2C.D4 you will need to describe the conditions that are usually used in the Haber process, and then describe what will happen to both the rate and the yield if the conditions are changed.

Lesson outcomes

You should:

- understand the concepts of atom economy and yield
- appreciate that the actual yield is less than the theoretical yield
- know about altering rates of reaction.

2.12 Our changing Earth

Get started

How do you think mountains and hills are formed? Discuss in groups different ways in which this could happen.

Key term

Plate tectonics – A theory which is based on the idea that the Earth's crust is divided into separate tectonic plates. These tectonic plates float and move around on the liquid mantle below the crust.

Mountain building

The Himalayas are the highest mountains in the world. Mount Everest is over 8800 metres high – that's more than 5 miles. But scientists have found fossils of sea creatures at the top of these mountains. This tells us that the rocks which make up the Himalayas were once at the bottom of a shallow sea. Scientists now know that this is because two massive slabs of rock, called **tectonic plates**, are being pushed together. Where the two plates meet, the rock crumples up and a mountain range is formed.

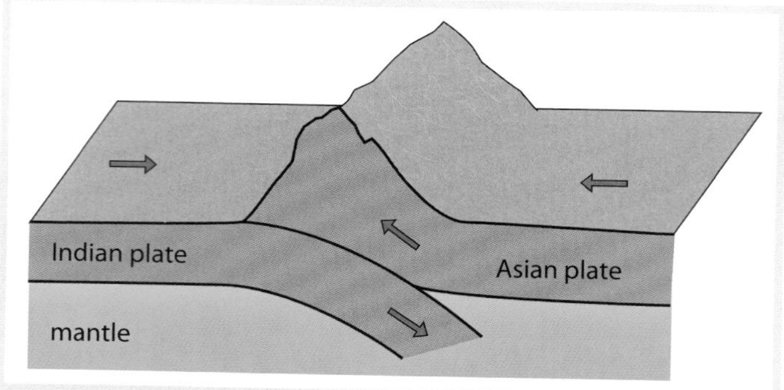

The formation of the Himalayas. The Indian plate is pushed below the Asian plate, and mountains form where the two plates meet.

Activity A

Several other mountain ranges were formed at plate boundaries. Research one of these mountain ranges and find out how it was formed. Include diagrams which show what happened where the two plates met. Possible mountain ranges to research include: the Andes and the Rocky Mountains.

Plate tectonics

The whole of the Earth's crust is divided into tectonic plates. You can see this on the map of the Earth's surface. These plates float on top of the **mantle** (the liquid rock under the Earth's crust). Huge swirling currents in the mantle make the plates move. The region where two plates meet is called a **plate boundary**.

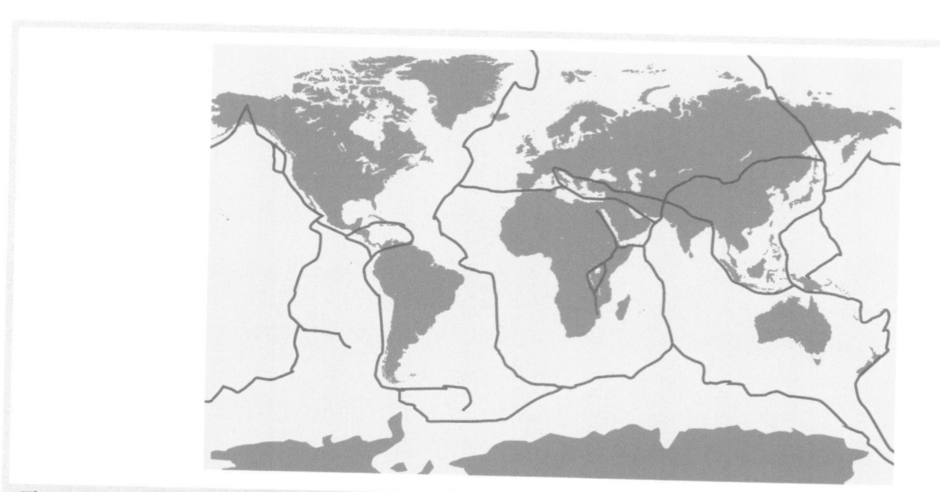

The tectonic plates (outlined in red) that make up the Earth's crust.

Did you know?

Some of Scotland's most famous landmarks are due to plate tectonics. Edinburgh castle stands on the rock of an old volcano, and Loch Ness is in a rift valley caused by a fault line.

Earthquakes and volcanoes

Plate tectonics also explain why some parts of the world are at risk from severe earthquakes or volcanic eruptions. Both of these events occur at plate boundaries. The plates could be moving apart, pushing together or sliding past each other.

Earthquakes When plates push together, earthquakes can happen because forces build up in the rock. Eventually the forces are too great and the rock shifts to release the pressure. If an earthquake happens under the sea it can trigger a **tsunami**.

Faults Huge blocks of rock can slip past each other causing cracks in the Earth's crust called **faults**. If you study satellite pictures or maps of the Earth you can pick out long straight valleys or sea inlets that are caused by the movement of rocks at faults.

Volcanoes Volcanoes are formed when **magma** (molten rock from beneath the Earth's crust) is forced up through the Earth's crust. We can see this rock erupting from the ground as molten **lava**. The lava can build up to form new mountains. Sometimes an eruption will blow a mountain apart, creating vast clouds of dust. Volcanoes also emit gases such as sulfur dioxide and carbon dioxide into the atmosphere.

A fault line thousands of miles long has created the Gulf of Aqaba (shown to the right of the peninsula) and the Dead Sea.

Activity B

Explain how rock movements could have created the Dead Sea and the Gulf of Aqaba, as shown in the satellite picture of the fault valley.

Assessment activity 2.7

| 2D.P9 | 2D.M7 (part)

You are a geologist who works for a research team that studies earthquakes and volcanic eruptions. Whenever there is an earthquake or volcanic eruption in any part of the world, members of the team fly out to study its effects.

You have been asked to prepare a poster about your work for a careers conference for Year 10 students at a local school. Your poster should be designed to motivate the students and also explain some of the science behind your work. You should explain:

- whereabouts in the world you might travel – use ideas about plate tectonics to suggest some possible locations for earthquakes or volcanic eruptions
- some of the effects you might observe in your job
- some of the things that have happened over the whole history of the Earth as a result of earthquakes or volcanic explosions.

Just checking

1 What is a tectonic plate?
2 Describe how mountain ranges can be formed by movements of tectonic plates.
3 State one way in which volcanoes can affect the Earth's atmosphere.
4 Name a feature on the Earth's surface that has been caused by movements of rocks at a fault line.

Did you know?

When a volcano called Mount Tambora, in Indonesia, erupted in 1815, the explosion converted 160 cubic kilometres of rock into dust. When the dust spread throughout the atmosphere it caused the global temperature to fall by more than 0.5 °C. This decrease was enough to cause unusually cool summers throughout the world resulting in terrible famines in the United States and Europe.

Tips

For 2D.P9 you need to describe how these natural factors have changed the Earth's surface and atmosphere. This could include recent changes as well as others that have happened over a much longer time span.

For 2D.M7 you should give some idea of how significant these changes are.

Lesson outcome

You should understand the natural activity factors like tectonic plates and volcanic eruptions that are affecting the Earth's crust.

2.13 Our changing atmosphere and oceans

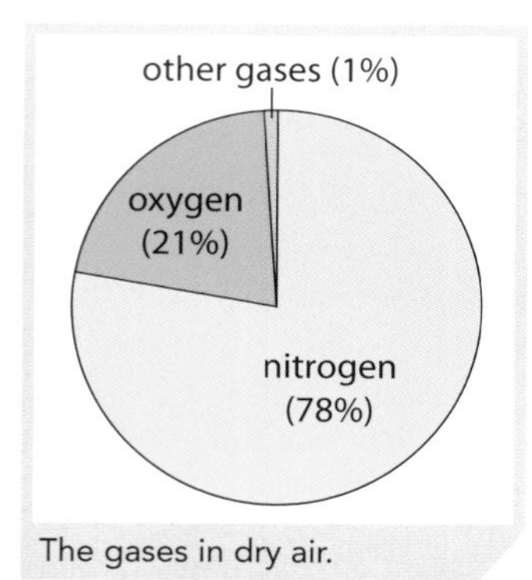

The gases in dry air.

The evolution of the atmosphere

Today, the air that we breathe is made up of about 78% nitrogen, 21% oxygen and 1% other gases, including water vapour and carbon dioxide.

Scientists think that the atmosphere was once very different to how it is today. Just after the Earth was formed, about 4.5 billion years ago, it may have had no atmosphere at all. Frequent volcanic eruptions released gases such as carbon dioxide (CO_2), water vapour (H_2O) and ammonia (NH_3) to form the early atmosphere. But this atmosphere changed over the next few billion years:

- water vapour condensed to form oceans
- some carbon dioxide dissolved in the oceans
- plants evolved and absorbed carbon dioxide converting it into oxygen by photosynthesis
- the ammonia reacted with oxygen to produce nitrogen.

So, by about 600 million years ago, these processes had produced an atmosphere of mostly nitrogen and oxygen. The amounts of oxygen, water vapour and carbon dioxide in the atmosphere have continued to change right up to the present day.

Early volcanoes released carbon dioxide and other gases into the Earth's atmosphere.

Changing atmosphere today

The changes which caused the atmosphere to evolve into today's atmosphere happened over billions of years. However, the concentrations of some gases in the Earth's atmosphere are changing much more rapidly than this.

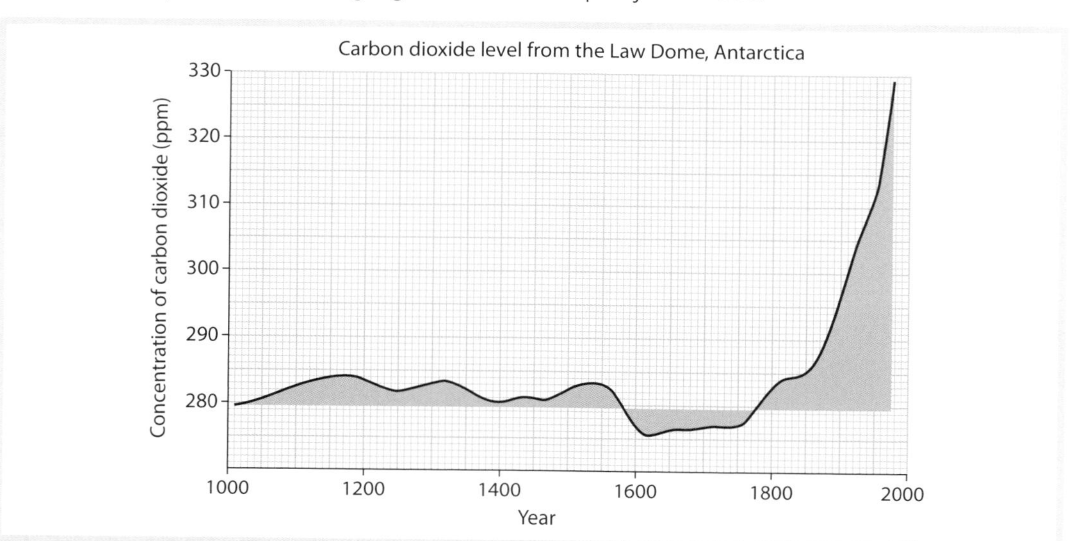

The amount of carbon dioxide in the Earth's atmosphere has increased very rapidly over the past 150 years.

Activity A

Find graphs that show how the oxygen content of the Earth's atmosphere has varied over the last billion years.

What do you notice? Do the same for carbon dioxide.

Take it further

The abbreviation 'ppm' stands for parts per million. It is used to give information about the concentration of substances when they are very low.
1 ppm = 0.0001%.

Scientists are worried by the rapid increase in the concentration of carbon dioxide in the atmosphere because it is a **greenhouse gas** and stops heat radiating away from the Earth. If less heat escapes, the overall temperature of the atmosphere rises. We call this process **global warming**.

Did you know?

Scientists have used computer models to make predictions about the effects of global warming produced by increases in the amount of greenhouse gases in the atmosphere. Even the most optimistic models predict an average temperature rise of more than 2 °C, but the warming will be much greater in the Arctic which may be more than 5 °C warmer. The melting ice from glaciers in Greenland could raise sea levels by more than 1 metre.

According to these models, the North Atlantic and the UK may be one of the regions of the world where the warming will be slowest.

Activity B

The concentration of CO_2 at the start of 2012 was 392 ppm.

1 Use the graph on the previous page to find the approximate concentration in the year 1700.

2 Estimate the likely CO_2 concentration in the year 2100 if concentrations go on rising at the same rate.

Evolution of the oceans

The oceans occupy 72% of the surface of the Earth. But how did the oceans form?

- The early Earth was very hot. Any liquid water would have evaporated.
- Some of the water vapour remained in the atmosphere and some was trapped in the molten rocks beneath the crust.
- Volcanoes released some of this trapped water vapour into the atmosphere.
- As the earth cooled down, the water vapour in the atmosphere condensed into clouds and fell to the earth as rain, creating the oceans.

The Pacific Ocean is the largest ocean on Earth. It is larger than all of the Earth's land area combined.

What has happened since then?

- Carbon dioxide from the atmosphere has dissolved in the oceans.
- Minerals and salts from rocks have dissolved in rain water and ended up in the oceans, making the oceans 'salty'.

Just checking

1 Carbon dioxide was one of the three main gases in the Earth's early atmosphere. Name the two other main gases.

2 In today's atmosphere there is much less carbon dioxide and a lot more oxygen. Explain why.

3 Carbon dioxide is a greenhouse gas. What does this mean?

Lesson outcome

You should understand the factors that influenced the evolution of the Earth's atmosphere and oceans.

2.14 The effects of human activity

Key term

Ore – A rock from which a metal can be extracted.

The Bingham Canyon is over a kilometre deep.

The biggest hole

The Bingham Canyon in Utah, USA, is the world's deepest man-made hole in the ground. It has been dug by a company that needs copper **ore** to use in its copper extraction plants. The Bingham Canyon supplies enough ore to produce 300 000 tonnes of copper every year – but the world demand for copper is increasing. This means there need to be many more mines like the Bingham Canyon.

Copper ore is a non-renewable resource. This means it is not being replaced in the Earth's crust. Scientists calculate that at the rate we are using up the copper ore it may begin to run out in about 60 years' time.

Life cycle of producing copper

Copper is needed for electrical wiring and water pipes, but there is an environmental cost in producing this copper.

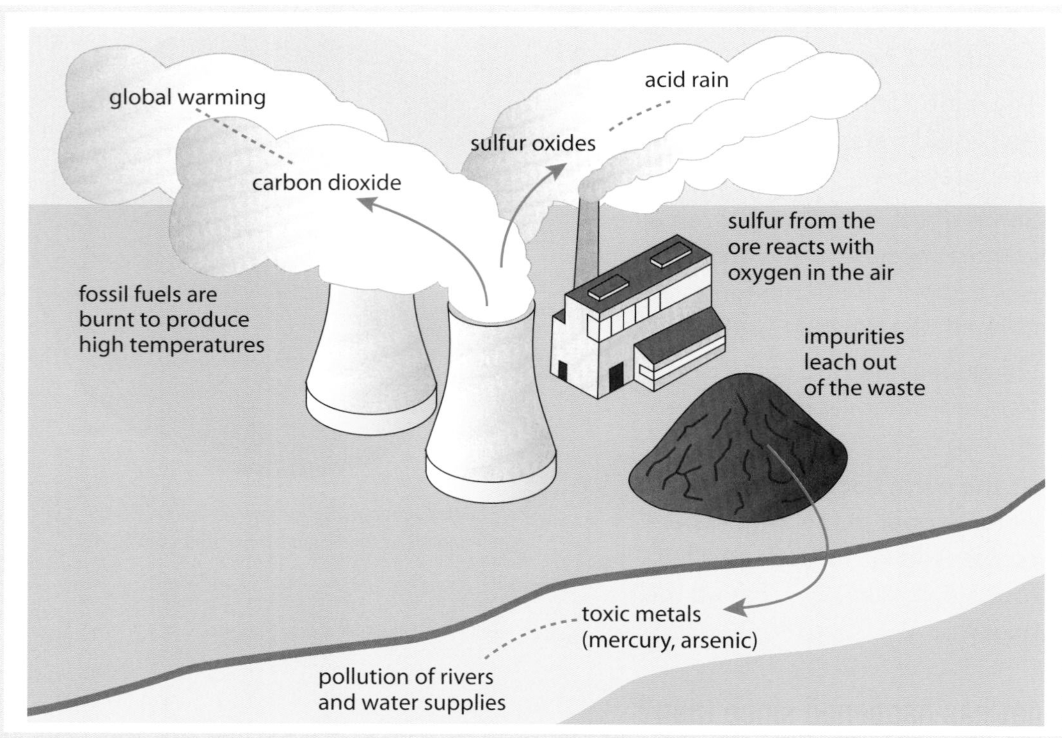

The environmental problems caused by the extraction of copper.

Safety and hazards

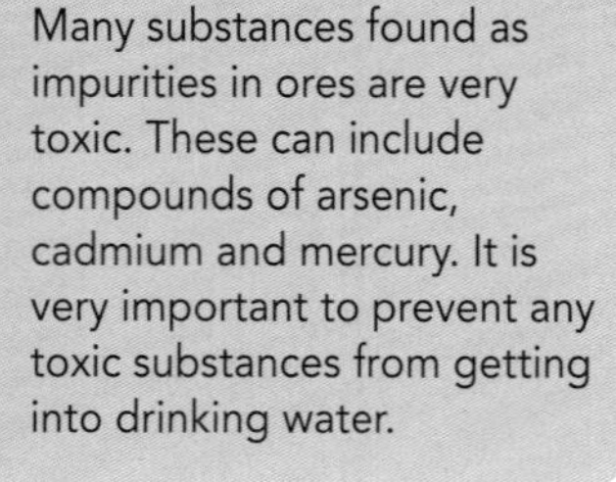

Many substances found as impurities in ores are very toxic. These can include compounds of arsenic, cadmium and mercury. It is very important to prevent any toxic substances from getting into drinking water.

In order for us to get a full picture of the environmental effect of manufacturing copper wire, we can do a life-cycle analysis.

- Obtaining the raw materials: digging metal ores out of the ground in open-cast mines. Left over material is disposed of in giant waste heaps.
- Producing a useful material: making pure copper from its ore uses energy. It also produces waste gases, like sulfur dioxide which causes acid rain, and tonnes of solid waste which can pollute rivers.
- Disposing of the material after it has been used: most copper wiring is recycled by melting it down. This also uses energy.

Activity A

Sodium chloride is needed to help soften water in dish-washers. Carry out a life-cycle analysis to help you consider the environmental effect of producing and using sodium chloride in this way.

Any process which uses energy will have an effect on the environment. This is because the energy usually comes from burning fossil fuels which releases carbon dioxide into the atmosphere. Carbon dioxide is a greenhouse gas which is thought to contribute to global warming. Burning some fossil fuels can also release sulfur dioxide.

Acid rain is formed when sulfur dioxide dissolves in rain water. When the rain water reaches the soil it can kill trees.

Who's affected?

Some human activities have effects within just a few miles of the activity (local effects), while others affect the whole Earth (global effects). The table shows that copper production has both local and global effects.

Activity	Effect
Obtaining the ore	Mining: local Disposing of waste: local Depleting the Earth's resources: global
Extracting copper from the ore	Acid rain: local (within a few hundred miles) Toxic waste: local
Energy used in these processes	Release of greenhouse gases: global

The production of copper is only one example of a human activity with environmental effects. Here are three others.

- Crude oil is a non-renewable resource. It is obtained from rocks underneath the surface of the Earth. It is transported in tankers to chemical refineries. Here, the different substances in crude oil are separated and processed to make fuels, plastics and other products.
- There are vast amounts of salt (sodium chloride) dissolved in sea water. In many countries, salt is obtained from sea water by using heat from the Sun to evaporate the water.
- The air contains nitrogen and oxygen gases. Air can be liquefied and then the different gases boiled off at different temperatures. This process requires energy from burning fossil fuels. It is important because we can use gases such as oxygen in hospitals.

Just checking

1 The raw materials from which chemists make useful products can come from the earth, the sea or the atmosphere. Give an example of a useful material that is obtained from **(a)** the earth, **(b)** the sea, **(c)** the atmosphere.

2 Name a harmful gas produced during the extraction of copper from its ore.

3 Environmental effects can be local or global. Give an example of **(a)** a local effect and **(b)** a global effect caused by the chemical industry.

Link

Acid rain is discussed in lessons 1.15 and 1.16.

For more about non-renewable resources, look at lesson 1.25.

Activity B

Petrol is a fuel made from compounds found in crude oil. List all the ways in which the production of petrol could cause environmental effects. Are these effects local or global?

Lesson outcome

You should understand how human activity factors affect the Earth and its environment.

2.15 Sustainable development – choices

Key terms

Sustainable – Describes ways in which human beings can meet their needs without exhausting the world's resources or polluting the environment for future generations.

Did you know?

Some companies are developing ways of capturing carbon dioxide before it is released into the atmosphere. The carbon dioxide is separated from the other gases given off when fossil fuels are burnt. It is then pumped underground or to the bottom of the sea where it is stored. The technology is expensive and it is difficult to store the gas securely.

Link

To find out more about how nuclear power is generated, see lesson 3.4.

Discussion point

Some people think that the UK should build more nuclear power plants to provide a greater fraction of our energy needs in the future. Do you agree? Discuss the arguments for and against this idea.

Lesson outcome

You should be aware of the issues and human choices involved in sustainable development; for example, recycling, use of fossil fuels versus nuclear fission fuels.

As you saw in the previous lesson, whenever a material is manufactured it uses up the Earth's resources and needs a source of energy. Society must decide how far it is prepared to go in making choices that could reduce the environmental impact of these processes. Some **sustainable** choices are given in the table below.

Possible choice	Arguments in favour	Arguments against
Recycle as much material as possible.	Less energy and raw materials are needed to make new items.	It is difficult to persuade everyone to recycle materials. Recycling can sometimes be expensive.
Use energy sources that do not use fossil fuels (e.g. nuclear power).	No carbon dioxide is produced from these sources. Fossil fuel resources are not used up.	People worry about the safety of nuclear power plants and the nuclear waste produced. Energy produced in this way may cost more than energy produced from fossil fuels.
Prevent waste products (e.g. SO_2) being released into the environment by using chemical scrubbers.	The waste products react to form safe and harmless products. The products are often useful and can be sold.	The technology could be expensive to install for developing countries.

In 2011, a huge earthquake and tsunami affected the Fukushima power plant in Japan, causing severe damage to three of the six nuclear reactors. People living within 20 km of the plant were evacuated.

WorkSpace

Isobel Richards

Environmental Impact Officer

I work for a large chemical company. We are developing new processes all the time. Every time we do this, we have to review the effect the process will have on the environment.

We are committed to making our industry as sustainable as possible – which means we need to reduce the amount of pollutants released and the resources we use, such as fossil fuels and raw materials. Modern chemical plants are very efficient, for example any heat produced gets recycled to heat other parts of the plant. New ways are being developed to prevent there being any waste products at all.

In the past, hazardous solvents like cyclohexane were used, but now we are developing safer 'green solvents'. For example, we use liquefied carbon dioxide, which could come from one of the new carbon capture plants being developed. My role is to show that over the whole life cycle of the process – from extraction to final disposal of material – the environment is not going to be damaged. Many of our new plants are built away from urban areas, but that means that I need to look at the effect they will have on the ecosystems which surround them.

In order to do my job I have to understand a lot about chemistry – but I also need a good grasp of ecology, economics and legal issues. It is very exciting to see what new technologies are being developed to change our industry from the old, dirty image we used to have. Now I reckon that our new plants are some of the safest and cleanest manufacturing units in the country.

Think about it

1 Isobel says that 'any heat produced in reactions gets recycled'. How do you think this is done? Try searching for 'heat exchanger' on the Internet.
2 What is a solvent? Find out about the hazards of cyclohexane solvents.
3 Suggest two reasons why liquid carbon dioxide is called a 'green' solvent.

2.16 Sustainable development – solutions

Key term

Renewable energy – Energy from natural sources that will never run out.

Many scientists believe that if we do not reduce emissions of carbon dioxide significantly over the next two decades then global warming will cause a temperature rise that will have catastrophic effects on the environment and the global economy. Scientists are developing new ways of reducing the use of fossil fuels in order to reduce carbon dioxide emissions. These are possible solutions.

- Make greater use of **renewable energy** sources, like wind and solar power.
- Use biofuels like **bioethanol**, which is produced from the sugar made in crops. When the fuel is burnt, carbon is released in the form of carbon dioxide. Since carbon is taken up from the atmosphere when the crop grows, the process is 'carbon neutral'. But this does not take into account the carbon dioxide produced when the crop is transported on trucks, or during its conversion to bioethanol.
- Develop completely new ways of generating energy, such as nuclear fusion. Unlike nuclear fission, this would not produce dangerous radioactive waste.

Link

See lesson 1.26 for more information about renewable energy sources.

You can read more about the process of nuclear fusion in lesson 3.6.

However, there are problems with all these methods.

Method	Problem(s)
Renewable energy sources (wind and solar power)	Equipment is expensive compared to the value of the energy generated. Some people do not like the look and noise of wind turbines and solar panels. They only work well in certain weather conditions.
Biofuels	Growing biofuels means that less land is available for food crops.
Nuclear fusion	A lot of expensive research is needed. It may take many decades to develop a working reactor. No-one knows whether it will eventually work or not.

Discussion point

The UK government has set a target of reducing emissions of CO_2 in the UK by at least 80% by 2050, relative to 1990 levels.

1 Discuss why the government has done this and whether you agree that it is a sensible policy.
2 What are the main changes that should be made in order to achieve this target?

Sugar from this plant can be used to manufacture ethanol, a biofuel.

Activity A

There have been several proposals recently to use the power of the tides in places like the Severn estuary as a source of renewable energy. Use the Internet to research how this could work and describe the advantages and disadvantages of the tidal schemes.

Assessment activity 2.8

| 2D.P8 | 2D.M7 | 2D.D5

1 You are a member of the public who is very concerned about a proposal to build a copper extraction plant near to your house. You are worried about the local effects of the process and also that it will have an impact on global warming. Write a letter in which you suggest alternative ways of meeting the demand for copper without building a new plant.

2 You are an environmental scientist working for the company developing the copper extraction plant. Respond to the member of the public about their concerns. You should explain some of the ways in which the plant could be designed to reduce its environmental impact.

Tips

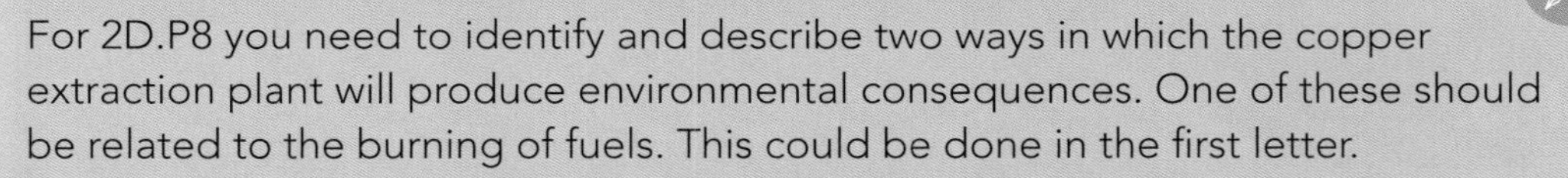

For 2D.P8 you need to identify and describe two ways in which the copper extraction plant will produce environmental consequences. One of these should be related to the burning of fuels. This could be done in the first letter.

For 2D.M7 you will need to make sure that in one of your letters you compare the effects which the plant will produce with effects produced by natural activities. You could refer back to some of the information you wrote about in the activities in 2.13.

For 2D.D5 you should describe and explain two different methods for reducing the environmental effects of the copper extraction plant and the burning of fossil fuels. You need to include an evaluation of how successful these methods are likely to be. This could be done in one or both of the letters.

Just checking

1 Nuclear fission is used as an energy source in many countries. Give one argument in favour of the use of nuclear fission and one argument against.

2 Name two sources of renewable energy.

3 Some people say that biofuels are 'carbon neutral'. What does this mean? Explain why biofuels may not be carbon neutral.

4 Why would a nuclear fusion reactor be safer than a nuclear fission reactor?

Lesson outcome

You should be aware of the human solutions that can lead to sustainable development, e.g. renewable energy, biofuels (ethanol), and nuclear fusion.

Introduction

Radioactivity can be extremely dangerous if it is not managed properly. Despite this, physicists have been able to make use of radioactive isotopes in many areas of our lives. In this unit, you will learn how radioactive materials are used in the home, in industry and in medicine.

Around the world, some countries use nuclear energy for generating electricity as the main alternative to power stations fired by coal and gas. You will learn about problems with using radioactive materials in nuclear reactors, including the impact of a nuclear accident.

In this unit you will also study how electricity is produced in batteries, solar cells and generators. You will learn how electricity is distributed to our homes and work places, and how wasting energy can be minimised.

Space, the last part of this unit, is a fascinating subject to study. You will learn how our own planet may have been formed, and you will study the evidence available to explain the origin of the Universe.

Assessment: You will be assessed using a series of internally assessed assignments.

Learning aims

After completing this unit you should:

a understand ionising radiation, its uses and sources

b know how electrical energy produced from different sources can be transferred through the National Grid to homes and industry

c know the components of the Solar System, the way the Universe is changing and the methods we use to explore space.

I really liked the radioactivity part of this unit, as it goes into more detail than we have covered before in school. It really showed me how useful radioactivity actually is! I also enjoyed learning about how our own Solar System was formed and how the Universe is changing all the time.

Tom, *16 years old*

Energy and Our Universe

3

This table shows you what you must do to achieve a Level 1 Pass, or a Level 2 Pass, Merit or Distinction grade, and where you can find activities in this book to help you.

Assessment criteria			
To achieve a Level 1 Pass grade, the evidence must show you are able to:	To achieve a Level 2 Pass grade, the evidence must show you are able to:	To achieve a Level 2 Merit grade, the evidence must show you are able to:	To achieve a Level 2 Distinction grade, the evidence must show you are able to:
Learning aim A: Understand ionising radiation, its uses and sources			
1A.1 Describe the structure of atomic nuclei. Assessment activity 3.1 Assessment activity 3.2	**2A.P1** Describe half-life in terms of radioactive decay. Assessment activity 3.2	**2A.M1** Use graphs to explain radioactive decay and half-life. Assessment activity 3.2	**2A.D1** Calculate the half-life of radioactive isotopes. Assessment activity 3.2
1A.2 Identify the types of ionising radiation. Assessment activity 3.1 Assessment activity 3.3	**2A.P2** Describe the different types of ionising radiation. Assessment activity 3.1 Assessment activity 3.3	**2A.M2** Compare the benefits and drawbacks of using radioactive isotopes in the home or workplace. Assessment activity 3.3	**2A.D2** Justify the selection of a radioactive isotope for a given use within the home or workplace. Assessment activity 3.3
1A.3 Identify the problems associated with the use of radioactive isotopes. Assessment activity 3.3	**2A.P3** Describe the problems associated with the use of radioactive isotopes. Assessment activity 3.3	**2A.M3** Describe the environmental impact of radioactive material from nuclear fission reactors released into the environment. Assessment activity 3.4	**2A.D3** Evaluate the environmental impacts of a nuclear fission reactor accident, in terms of half-life. Assessment activity 3.4
1A.4 Describe nuclear fission and fusion. Assessment activity 3.4 (part) Assessment activity 3.5 (part)	**2A.P4** Describe how controllable nuclear fission and fusion reactions are. Assessment activity 3.4 (part) Assessment activity 3.5 (part)		
Learning aim B: Know how electrical energy produced from different sources can be transferred through the National Grid to homes and industry			
1B.5 Identify methods of producing electricity from different sources. Assessment activity 3.7	**2B.P5** Describe methods of producing a.c. and d.c. electricity. Assessment activity 3.7	**2B.M4** Compare the efficiency and environmental impact of electricity generated by different sources. Assessment activity 3.7	**2B.D4** Assess, in quantitative terms, ways to minimise energy losses either when transmitting electricity or when transforming electricity into other forms for consumer applications. Assessment activity 3.8
1B.6 Demonstrate building simple series and parallel circuits. Assessment activity 3.6	**2B.P6** Use $V = IR$ to predict values in electric circuit investigations. Assessment activity 3.6	**2B.M5** Assess, in qualitative terms, ways to minimise energy losses when transmitting electricity. Assessment activity 3.8	
1B.7 Describe electrical power in terms of voltage and current. Assessment activity 3.8	**2B.P7** Describe how electricity is transmitted to the home or industry. Assessment activity 3.8		

Assessment criteria			
Learning aim C: Know the components of the Solar System, the way the Universe is changing and the methods we use to explore space			
1C.8 Identify the components of our Solar System. Assessment activity 3.9	**2C.P8** Describe the structure of the Universe and our Solar System. Assessment activity 3.9 (part) Assessment activity 3.11 (part)	**2C.M6** Describe how the Universe and the Solar System were formed. Assessment activity 3.13	**2C.D5** Evaluate the evidence leading to the Big Bang theory of how the Universe was formed. Assessment activity 3.13
1C.9 Identify methods of observing the Universe. Assessment activity 3.10	**2C.P9** Describe the suitability of different methods for observing the Universe. Assessment activity 3.10	**2C.M7** Explain how evidence shows that the Universe is changing. Assessment activity 3.12	
1C.10 Describe the dynamic nature of our Solar System and Universe. Assessment activity 3.12	**2C.P10** Identify evidence that shows the dynamic nature of the Universe. Assessment activity 3.12		

How you will be assessed

The unit will be assessed by a series of internally assessed tasks. You will be expected to show an understanding of physics in the context of the nuclear, electrical and astronomy sectors. The tasks will be based on numerous scenarios which place you as the learner in the position of working in a number of industrial sectors; for example, in the medical physics department of a hospital, as a nuclear technician and an electrical technician, and as an assistant working for the national space centre.

Your actual assessment could be in the form of:

- a written observation of experiments you will carry out
- training materials, such as leaflets and presentations
- tables and posters of results of investigations.

3.1 Ionising radiation

Key terms

Radioactive – Releases radiation.
Mass (nucleon) number – Number of protons and neutrons in the nucleus of an atom.
Atomic (proton) number – Number of protons in the nucleus of an atom.
Isotopes – Atoms that have the same number of protons but different numbers of neutrons.
Ion – An atom which has a charge such as +2.

Link

Look at lesson 1.10 to remind yourself about atomic structure.

Look at lesson 1.11 to remind yourself about isotopes and about the importance of carbon-14 for carbon dating.

See lesson 3.3 for more on alpha, beta and gamma decay.

Did you know?

Radioactivity was discovered by accident by the French physicist Henri Becquerel in 1896. The becquerel is one of the units used to measure the radioactivity of a material.

Nuclear structure

Why are some materials **radioactive**? You will know from Unit 1 that atoms have electrons that orbit the nucleus. Inside the nucleus are protons and neutrons. The total number of neutrons and protons in a nucleus is called the **mass number** or **nucleon number**. The number of protons is called the **atomic number** or **proton number**. The mass number and atomic number can be shown like this:

mass number → $^{12}_{6}C$
atomic number →

Atoms of an element that have the same number of protons but different numbers of neutrons are called **isotopes**.

- Carbon-12 is the most common isotope of carbon. Each atom has six protons and six neutrons in the nucleus. The nuclear symbol is $^{12}_{6}C$.
- Carbon-14 ($^{14}_{6}C$) is a radioactive isotope of carbon. The nucleus of each atom has six protons and eight neutrons.

Types of radiation

Some isotopes are unstable, and **decay** by emitting radiation. There are three types of radiation: alpha (α), beta (β) and gamma (γ).

- In alpha radiation atoms emit alpha particles, which consist of two neutrons and two protons. This is the same as in the nucleus of a helium atom. Alpha particles have a charge of +2.
- Beta particles are electrons that are emitted from the nucleus of atoms. They have a charge of −1.
- Gamma rays are a form of electromagnetic radiation and do not have a charge.

Radiation type	What it is/charge	Ionisation	Penetration power
Alpha (α)	2 protons and 2 neutrons Charge: +2	Highly ionising	Weakly penetrating. Stopped by a sheet of paper (or human skin), but very dangerous if swallowed.
Beta (β)	Electron from nucleus Charge: −1	Moderately ionising	Moderately penetrating. Stopped by 1 mm of aluminium.
Gamma (γ)	Electromagnetic waves Charge: neutral	Weakly ionising	Highly penetrating. Stopped by 10 cm of lead.

Ionising radiation

Alpha particles, beta particles and gamma rays are all forms of ionising radiation. This means the radiation can cause atoms to gain or lose electrons and form **ions**.

Alpha particles are large and heavy (they have the most mass), so they are very ionising. However they cannot penetrate very far through materials. Gamma rays are only weakly ionising, but they pass through many materials easily. Beta particles are somewhere in between.

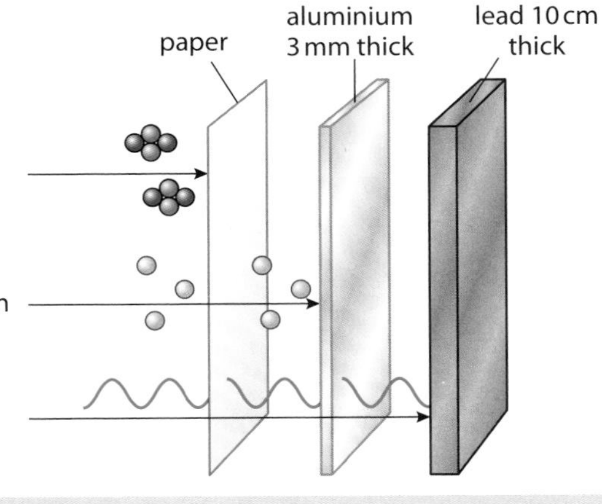

The penetrating power of different types of radiation.

Background radiation

There are many everyday things that are radioactive. Just eating, drinking and breathing exposes us to radiation. You can be exposed to radiation by entering any hospital that uses X-rays. In certain areas of the country the rocks and soil contain radioactive elements. Rocks containing uranium are radioactive themselves; they also produce radioactive radon gas. This can collect inside houses unless they are ventilated properly. Radon is the biggest source of background radiation.

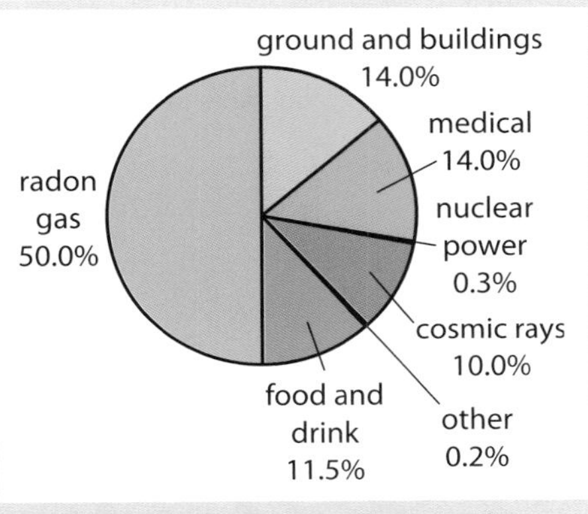

Pie chart showing the main sources of background radiation.

Assessment activity 3.1

2A.P2

1 You are a junior journalist working on a science magazine. The editor has written an article about smoke alarms, which rely on the isotope americium-241.
 (a) Americium-241 has 146 neutrons. What is its proton number? What is its nucleon number?
 (b) Write the nuclear structure in symbols.
 (c) Describe the type of radiation americium-241 emits.
2 The editor wants to write a feature box called 'What is radiation?' to go with her article, but her physics isn't very good. Write some notes to help her. Explain:
 (a) what an isotope is
 (b) why isotopes such as americium-241 are radioactive
 (c) the three types of radiation produced by radioactive materials.

Just checking

1 What are the differences between alpha, beta and gamma radiation?
2 Rank the different types of radiation from most penetrating to least penetrating.
3 Rank the different types of radiation from most ionising to least ionising. Explain your answers.

Activity A

An environmental health technician uses a Geiger–Müller (GM) tube to measure the amount of radiation emitted by two unknown substances. She observes the following.

Sample A registers a large signal by the GM tube, with nothing placed in front of it. The signal remains high even when a block of aluminium is placed in front of it.

Sample B registers a large signal when a 5-cm piece of wood is placed in front of it, but registers almost zero when a block of aluminium is placed in front of it.

1 What types of radiation are emitted by substance A and by substance B? Explain your answer.
2 Investigate where in the UK you are likely to find radon under houses.

Tip

For 2A.P2, it is not enough to simply give the names of the different types of radiation. You need to describe what each type of radiation is, and what its properties are.

Lesson outcomes

You should know how to write down the structure of nuclei and be able to identify the three types of ionising radiation.

3.2 Radioactive decay and half-life

Key terms

Radioactive decay – When the nucleus of an unstable isotope changes by giving off radiation.

Random – We can't say which particular nucleus is going to decay at a particular time.

Half-life – The time it takes for half the nuclei in a sample of a radioactive isotope to decay.

To find out why some isotopes are radioactive and others are stable, we have to look inside the nucleus.

In many atoms, the forces in the nucleus are balanced and the nucleus is stable. However, in some atoms the forces in the nucleus are unbalanced, or the nucleus has too many neutrons or protons. These atoms are unstable. It is this instability that causes the atoms to be radioactive and emit radiation. This is called **radioactive decay.**

Radioactive decay is a **random** process. This means that we cannot predict when any particular atom will decay. However, we can predict how long it will take for half of the radioactive atoms in a sample to decay. This is the **half-life**.

The number of nuclei that decay each second is the activity of the sample. The activity can be measured in counts per second, or **becquerels (Bq)**, using a Geiger–Müller tube. As the sample decays, the number of undecayed nuclei goes down and the activity decreases. Another way to measure half-life is to measure the time taken for the activity of a material to go down to half its original value.

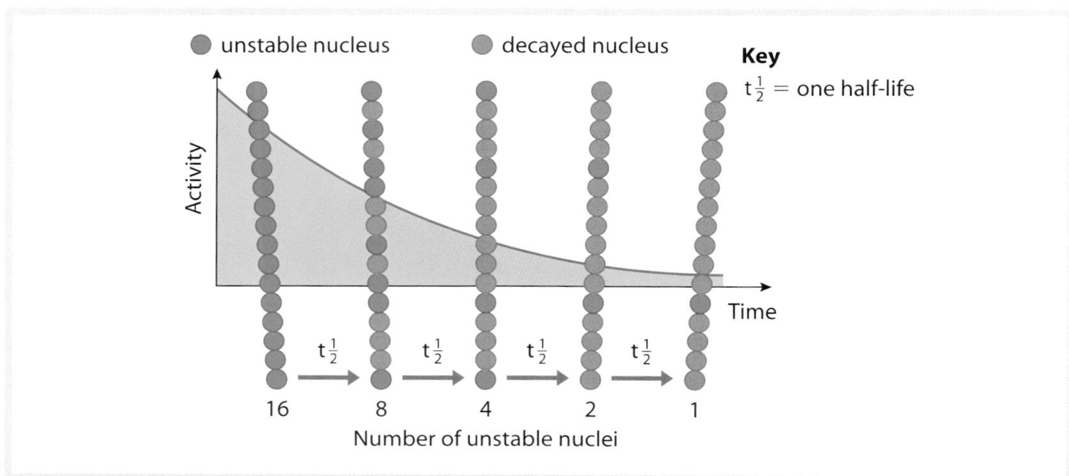

A graph of activity versus time for radioactive decay. The diagram also shows how with each half-life the number of undecayed nuclei halves.

Calculating half-life

The graph below shows how activity changes with time for a radioactive source.

The activity at 5 minutes is 600 Bq. The activity is 300 Bq at 12 minutes. The half-life of this isotope is 7 minutes (12 – 5 = 7).

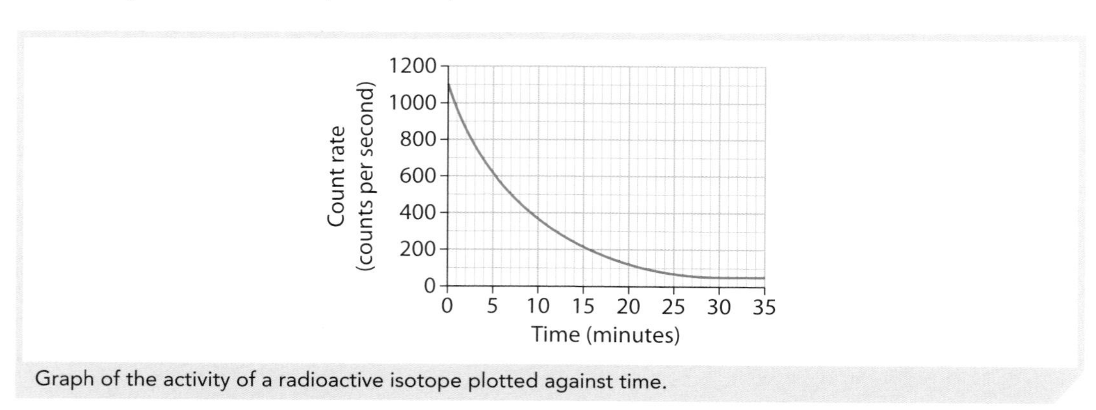

Graph of the activity of a radioactive isotope plotted against time.

Activity A

Investigate the random nature of radioactivity. You will need about 100 coins to simulate atoms. 'Tails' represents undecayed atoms, 'heads' represents atoms that have decayed. Place the coins in a box, shake them well, and pour them out on to the floor. Count all the coins that are heads, and put them aside. Record the result in a table. Put the 'undecayed atoms' (those that were tails) back in the box and shake again. Repeat the procedure until all the atoms have decayed (come up heads).

1 Plot a graph of the number of coins that have decayed at each throw (the count) against the throw number.
2 How does your graph compare with the graph of the radioactive isotope strontium-90? (Hint: you will have to look up the graph for strontium-90.)
3 From your graph, determine how many throws are needed for 7/8 of the coins to 'decay'.

Worked examples

1 Iodine-131 has a half-life of about 8 days. A sample of iodine-131 has a mass of 50 mg. Calculate how much I-131 will be left after **(a)** 8 days, **(b)** 24 days.
   (a) 8 days is the half-life, so 25 mg will be left.
   (b) 24 days is three half-lives, so 6.25 mg will be left: 50 → 25 → 12.5 → 6.25.
2 The activity of a radioactive isotope is 1000 Bq. 6 hours later the activity is 125 Bq. **(a)** What is the half-life? **(b)** What fraction of the original isotope will be left?
   (a) 6 hours is three half-lives: 1000 → 500 → 250 → 125
   Half-life = 2 hours.
   (b) One eighth of the original material will be left: $1 \rightarrow \frac{1}{2} \rightarrow \frac{1}{4} \rightarrow \frac{1}{8}$.

Assessment activity 3.2

| 2A.P1 | 2A.M1 | 2A.D1

A nuclear technician carried out two experiments on different radioactive substances, as detailed below.

Experiment 1: The table shows data for the radioactive isotope polonium-218. Polonium-218 decays by emitting an alpha particle.

Time (min)	0	1	2	3	4	5	6	7
Count rate (counts per second)	550	405	320	258	204	164	128	102

(a) Study a copy of the periodic table and write down the nuclear symbol for polonium-218.
(b) Write down what is meant by half-life.
(c) Plot a graph of the count rate against time.
(d) Use the graph to determine the half-life of polonium-218.

Experiment 2: A mystery radioactive substance registered a count rate of 3600 per second at the start of the experiment. Ten minutes later it registered 1800 counts per second.
(a) What is the half-life of the mystery material?
(b) How long would it take for the count rate to go down to 450 particles a second?

Tips

For 2A.P1, you need to refer to 'nuclei' not 'an atom' in the description of half-life.

For 2A.M1 you need to refer to the shape of the radioactive decay curve and the fact that the decay is a random process.

2A.D1 can be achieved by making sure that you show your workings out and give the answer in the correct units.

Lesson outcomes

You should:
- understand that radioactive decay is a random process
- be able to calculate half-life.

3.3 Uses of ionising radiation

Ionising radiation

Different radioactive isotopes emit different types of radiation. This means that they can be used in a wide variety of ways.

Did you know?

Before the hazards of radioactive isotopes were properly understood, the radioactive element radium was used to make luminous hands for clocks and watches.

Alpha particles

Alpha radiation is strongly ionising. This makes isotopes producing alpha radiation useful in devices called thermoelectric generators, which produce electricity using a source of alpha radiation.

Alpha radiation is also useful in smoke detectors. It is safe to use in a smoke detector as alpha particles are stopped by a few centimetres of air or by the skin. However, tampering with a smoke alarm could be dangerous if the isotope is breathed in or swallowed. This is because at short distances alpha particles are highly ionising and damage cells.

The Cassini–Huygens space probe was launched in 1997 and has been studying Saturn and its moons since 2004. It gets its electricity from a thermoelectric generator powered by plutonium-238, which is a powerful alpha emitter.

Beta particles

Beta radiation can penetrate small thicknesses of material, so sources that produce beta radiation are used, for instance, to measure the thickness of paper. The isotope is placed on one side of a moving sheet of a material and a GM tube on the other. The count rate changes as the thickness of the sheet varies. If a gamma ray emitter was used, the GM tube would not show any differences in the level of radioactivity when the thickness of the paper changed. This is because almost all of the gamma rays would just pass straight through the material being measured.

Sources of beta radiation are also used to treat some types of cancer. The source is located close to the cancer cells so that the radiation damages these cells but not healthy cells nearby.

Gamma rays

Gamma rays can travel through many types of material. This makes them useful as medical tracers for patients with a defective thyroid gland. A liquid containing radioactive iodine is given to the patient either in a drink or by injection. The iodine concentrates in the thyroid, and the gamma rays it emits can be used to form an image. The gamma rays emitted by ^{131}I pass through the body and are detected by a gamma camera. However the beta particles may be absorbed by body tissues, causing damage to healthy cells.

Source	Properties	Uses
Americium-241	• Alpha particle emitter • Half-life of 432 years – it will emit radiation over a long time	• Used in smoke alarms
Cobalt-60	• Gamma ray emitter • Half-life of 5.2 years	• Sterilising medical equipment • Irradiation of food to keep it fresh for longer • Treatment of some cancers • Used to check for flaws in welding or other structural elements
Iodine-131	• Beta particle and gamma ray emitter • Half-life of about 8 days – radioactivity falls to low levels relatively quickly	• Tracer in patients with defective thyroid gland
Strontium-90	• Beta particle emitter • Half-life of over 28 years – source would not need to be replaced very often.	• Measuring the thickness of materials such as paper or plastic
X-ray	• X-rays are similar to gamma rays, but they are manufactured rather than naturally occurring	• Medical imaging – for example of the skeletal system or in computed tomography (CT) scans • In industry, for detection of faults in structures • Also used for airport security, chemical analysis and astronomy

Just checking

1 Explain why isotopes like americium-241 are only a small risk to human health.
2 Describe some of the effects of radiation on living cells.

Safety and hazards

A radioactive hazard warning sign.

Hazards and safety measures

Ionising radiation can be bad for human health. Radiation can damage or kill living cells. Exposure to large doses can cause radiation sickness, leading to symptoms such as radiation burns, nausea, and even death.

People exposed to small amounts of ionising radiation over time have a high risk of suffering from cancer. Children born to parents who have been exposed to radiation are at risk of being born with a genetic disorder.

Despite these risks, radioactive isotopes can be used safely. Both alpha and beta radiation can be easily shielded so that the radiation does not spread far. In medicine, the isotopes used in people's bodies have a short half-life. They produce only a small amount of radiation before they decay.

Did you know?

People who regularly use equipment such as X-ray machines work from a shielded area, so that they are not exposed to the radiation. Patients who have X-rays receive only a small dose of radiation, not enough to be harmful.

Activity A

A hospital has asked you to produce an information leaflet for patients and their families about radioactive tracers. In your leaflet you should explain:

(a) why radioactive tracers are useful

(b) why it is safe for patients to have tracers inside their body, even though they are radioactive.

Assessment activity 3.3

| 2A.P2 | 2A.P3 | 2A.M2 | 2A.D2

You have just been appointed as a government technical advisor on the use of radioactive isotopes. You are required to produce a leaflet that will give a better understanding of the benefits and problems involved in using radioactive isotopes in the home and workplace. Include the following points when you produce your leaflet.

1. Identify and describe the different types of ionising radiation, using diagrams to explain your answers.
2. Identify some of the problems with using radioactive isotopes. List some of the safety measures that help to protect against these problems.
3. Compare the drawbacks of using radioactive isotopes with the benefits of their use.
4. Select one radioactive isotope that is used in the home or workplace and give reasons why this isotope is used for the particular application.

Tips

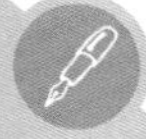

Task 1 provides an opportunity to meet 2A.P2. Make sure that your diagrams are labelled. For 2A.P3 you must describe the problems with alpha, beta and gamma radiation.

For 2A.M2 you need to make sure that you *compare* and not just describe the advantages and disadvantages.

For 2A.D2 you need to be very specific in the reasons for using an isotope. Make a note of the following:

- What is the isotope?
- What is its half-life and why is that important?
- What radiation does it emit?
- How is the isotope used in a specific application?

Lesson outcome

You should be able to describe the benefits and problems of using different radioactive isotopes.

WorkSpace

Mohammad Khan

Radiotherapy physicist

I work as a radiotherapy physicist in one of the national cancer centres.

Radiotherapy is the treatment of disease using X-rays, gamma rays or other ionising radiation. The ionising radiation damages the DNA of cells that are rapidly dividing, such as cancer cells, so it can stop the growth of tumours. Patients can be treated either with a beam of radiation, or with a small amount of radioactive isotope inserted directly into the tissue using a needle or wire.

My job has three main parts.

The first is to carry out regular checks on the equipment used to treat patients. I have to make sure that everything is working to a very high standard.

The second part of my job is to plan the treatment that a patient is going to receive. For example, I make sure that they are given a dose of radiation that is high enough to kill cancer tissue but causes the minimum of damage to healthy tissue. I work closely with doctors so we can correctly interpret the data following the treatment.

My job also involves using computer models to investigate how radiation interacts with human tissue.

To be successful in this role, I have to work well with other professionals. For example, I work with material scientists and electronic engineers to help design new materials and devices for use in medical treatment.

Think about it

1. Why do you think regular testing is an important part of a radiotherapy physicist's job?
2. As well as the dose level, what other important information will Mohammad need to know about the ionising radiation that is being used?
3. Mohammad's job also involves modelling how radiation behaves in the body. Why is this important?
4. As well as the physics of radiation, what other knowledge of physics will Mohammad need to have?

3.4 Nuclear fission

Get started

It is known that energy is given off during a nuclear reaction. In groups, discuss where you think the energy comes from. Use scientific terminology in your discussions.

Key terms

Nuclear fission – Large nuclei breaking down to form small nuclei.

Control rods – These are rods, usually made of boron, that absorb neutrons to control the speed of the reaction or stop the fission reaction altogether.

Fuel rods – These are rods that contain the material, like uranium, that splits during the nuclear fission reaction.

Radioactive decay is not the only kind of nuclear reaction. Some nuclei can be split apart by being bombarded with neutrons. When a neutron hits a nucleus, the nucleus splits into two almost equal parts. This is called **nuclear fission**. Fission reactions release a lot of energy.

- The process of nuclear fission starts when a large, unstable nucleus, such as uranium-235, absorbs a neutron.
- This makes the nucleus unstable and it splits into two smaller nuclei, called **daughter nuclei**. The daughter nuclei for uranium-235 are krypton-91 and barium-142. Three more neutrons are also released, plus a lot of energy.
- These fast-moving neutrons can then be absorbed by other uranium nuclei which can split and release more neutrons.
- This process is called a **chain reaction** and more energy is given off at each stage.

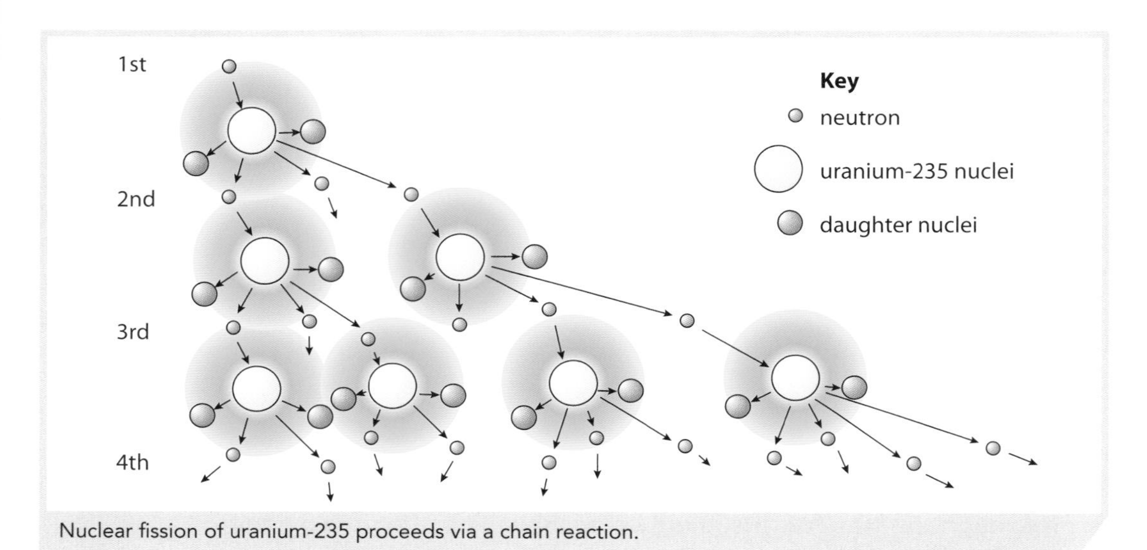

Nuclear fission of uranium-235 proceeds via a chain reaction.

Nuclear reactors

Once a chain reaction begins, it can easily get out of control. A runaway chain reaction could produce a nuclear explosion. This is how nuclear weapons work.

In a nuclear reactor, the chain reaction is controlled very carefully. The reactor produces heat, which is used to generate electricity. The diagram on p. 131 shows the inside of a reactor.

Did you know?

The very first nuclear reactor, designed by Italian scientist Enrico Fermi, was built in 1942. It was a pile of blocks of uranium and graphite (the material used for the 'lead' in pencils). The uranium was the fuel, while the graphite acted as a moderator. It was built in an old sports centre in Chicago, USA.

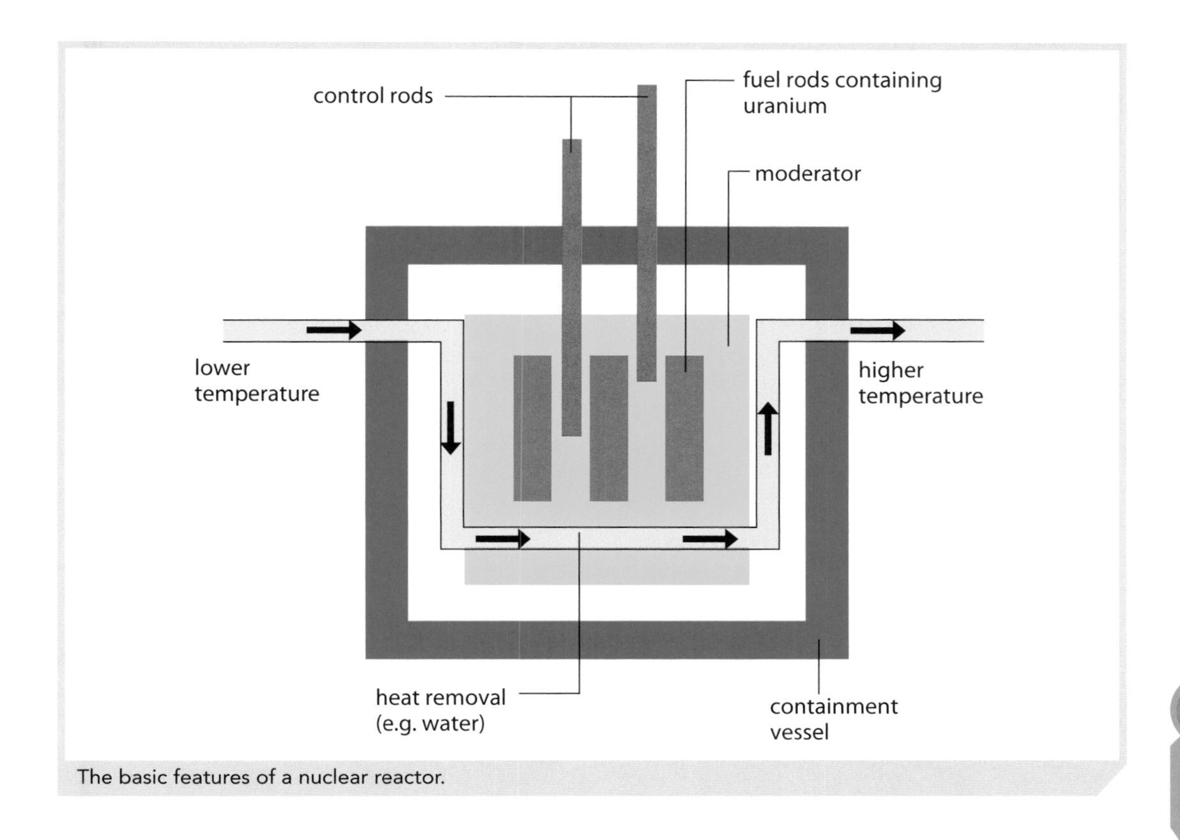

The basic features of a nuclear reactor.

Link

You learned about how a nuclear power station produces electricity from steam in lesson 1.21.

- **Fuel rods**: the fuel in many nuclear reactors is uranium-235. Naturally occurring uranium does not contain much of this isotope, so nuclear fuel is enriched to contain more U-235.
- **Moderator**: the fission reaction produces fast-moving neutrons, but slow-moving neutrons are much better at splitting uranium nuclei. The moderator is a material such as graphite, that slows down neutrons.
- **Control rods**: these are usually made of boron, a material that absorbs neutrons. They can be raised and lowered to manage the speed of the reaction. If they are lowered fully, they absorb the neutrons produced by fission and stop the chain reaction.
- **Coolant**: a coolant (often water) carries away a lot of the heat produced by fission. This energy is used to produce electricity. By cooling the reactor, the coolant also helps to control the chain reaction.
- **Reactor containment**: the whole reactor is encased in a thick containment vessel. This is the outer barrier to stop radiation from leaking. In an emergency it can contain excess heat and gases for a short period.

Activity A

1. Explain how the fission process in a nuclear reactor is controlled.
2. Using the Internet or national newspapers, find out how many nuclear power stations are in use in the UK. How does this compare to France?

Lesson outcomes

You should be able to describe nuclear fission and how it can be controlled in a nuclear reactor.

3.5 Safety in nuclear reactors

Key terms

Spent fuel – This is the fuel that is left over after the nuclear fission reaction has finished.

Fallout – Radioactive dust and ash that is carried through the air after a nuclear accident and falls to Earth far from the accident site.

The safety of nuclear reactors is very important. The factors to consider are:

- how to prevent the release of radioactive material
- how to store the waste products safely.

Preventing release of radioactive material

- The containment vessel housing the reactor is made out of thick steel and shielded by thick concrete walls so that any beta particles, gamma rays or neutrons that escape can be absorbed.
- If there is a problem and the reactor gets too hot, an emergency shut-down system automatically inserts the control rods fully into the core. This absorbs the neutrons and stops fission taking place.

Storing nuclear waste

Spent fuel is still highly radioactive, and also hot. Some of the radioactive isotopes in this waste material have half-lives of thousands of years. The waste therefore needs to be stored in a way that will be secure for a very long time.

- In the first stage of the process, the waste is stored in large ponds of water, where it cools down. Waste is stored in this way for several years.
- In some countries the fuel rods are then reprocessed, to remove any remaining useful fuel. Not all countries reprocess fuel.
- Reprocessed fuel is encased in glass and stored in stainless steel canisters.
- Most countries have plans for long-term storage of waste fuel in sealed concrete containers deep underground. Some nuclear waste storage facilities are being built, but no wastes are yet stored at these sites.

Cooling ponds, where spent fuel is cooled. These ponds are extremely hot and highly radioactive.

Nuclear accidents

Fukushima

Although nuclear power stations are very safe, there have been a few serious nuclear accidents.

In March 2011, a **tsunami** resulted in damage to the Fukushima Daiichi nuclear reactors in Japan. Investigations revealed the following effects of the damage.

- Traces of radioactive iodine-131 and caesium-137 were detected in Hawaii, Alaska and Montreal. Cs-137 has a half-life of about 30 years, and so it will be about 150 years before levels become insignificant.
- Radioactive iodine-131 was found in water purification plants in Tokyo (200 km from Fukushima). I-131 has a half-life of 8 days, so its effects were immediate but short-lived. In humans I-131 becomes concentrated in the thyroid gland leading to an increased risk of thyroid cancer.
- Radioactive traces were detected in vegetables, meat, fish and milk up to 300 km from Fukushima.
- Isotopes of strontium and plutonium were detected in the atmosphere, possibly due to contaminated steam escaping from the broken containment vessels.

Chernobyl

In 1986, there was a major nuclear accident in Chernobyl, Ukraine. There was an explosion in the reactor, and radioactive dust and ash was detected around the world. This nuclear accident resulted in the immediate death of nearly 30 people, some 300 people suffered from radiation sickness, and there was an increased risk of people developing cancer. Around 100 000 people were evacuated immediately following the disaster. Even today, there is an exclusion zone around the site.

Fallout from the Chernobyl accident had effects across Europe. In Britain, the biggest effects were on farms in Wales. The grassland became contaminated with caesium-137, and this affected the sheep grazing there. Even now, over 25 years after the accident, there are restrictions on moving and selling sheep from affected areas.

Activity A

The UK is one of the countries where nuclear waste is reprocessed. Some people are strongly opposed to the process. Find out about nuclear reprocessing from books and the Internet, then answer these questions.

1 Where is nuclear waste reprocessed in the UK?
2 What are the advantages of reprocessing fuel?
3 What are the disadvantages?
4 On balance, do you think that reprocessing is a good thing or a bad thing?

Assessment activity 3.4

| 2A.P4 (part) | 2A.M3 | 2A.D3

1 Using diagrams, describe nuclear fission.
2 Using a nuclear reactor as an example, explain how nuclear fission can be controlled.
3 Using the Japanese nuclear disaster as your example, describe the environmental impact when radioactive isotopes are released.
4 Explain why the Japanese government might be more worried about levels of caesium-137 than iodine-131.

In May 2012, bluefin tuna caught off the coast of California, USA were found to contain radioactive materials from the Fukushima nuclear accident. Scientists were surprised at how quickly the radioactivity had spread. The tuna were still safe to eat.

Did you know?

The radiation released during the Chernobyl nuclear disaster was nearly 400 times more than that caused by the nuclear bomb that hit Hiroshima, Japan, in the Second World War.

Tips

For 2A.P4 make sure you label the nuclear reactor and explain how each feature allows nuclear fission to be controlled. Tasks 3 and 4 for 2A.M3 and 2A.D3 can be attempted together.

Lesson outcome

You should know about the environmental issues associated with nuclear reactors.

3.6 Nuclear fusion

Key terms

Nuclear fusion – Nuclear reaction in which nuclei fuse together to form a bigger nucleus, releasing lots of energy. The energy released in a fusion reaction is greater than that from a fission reaction.

Nuclear fusion is the most important source of energy in the Universe. Without fusion we wouldn't exist, nor would the Sun or any of the stars.

In nuclear fusion, two small atomic nuclei join together to make a larger one. Fusion only takes place at very high temperatures and pressures, because it is hard to get nuclei close enough to join. When fusion does happen, even more energy is released than during fission.

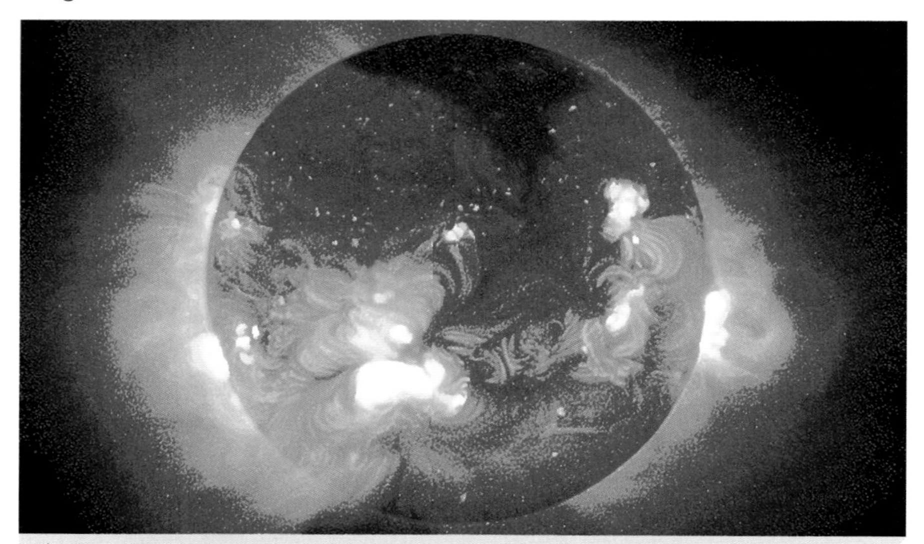

This X-ray image gives a sense of the enormous amount of energy the Sun produces. The whole surface boils and swirls, and huge flares burst out into space.

Fusion in the Sun

The Sun and the stars are all massive fusion reactors. The Sun is an enormous ball of hydrogen gas. At the Sun's core, hydrogen nuclei crash together to form larger helium nuclei, releasing energy in the process.

A hydrogen nucleus is a single, positively charged proton. Objects with the same charge repel one another, so it takes a lot of energy to make two hydrogen nuclei fuse. The temperature of the Sun's core is around 15 million °C and the pressure is about 2½ billion times Earth's atmospheric pressure. These very high temperatures and pressures provide the energy needed for fusion to take place.

Did you know?

To achieve the high temperatures and pressures needed for nuclear fusion, scientists have designed the ITER (International Thermonuclear Experimental Reactor) which will provide a temperature of 200 million °C. This is about 10 times higher than the temperature at the core of the Sun.

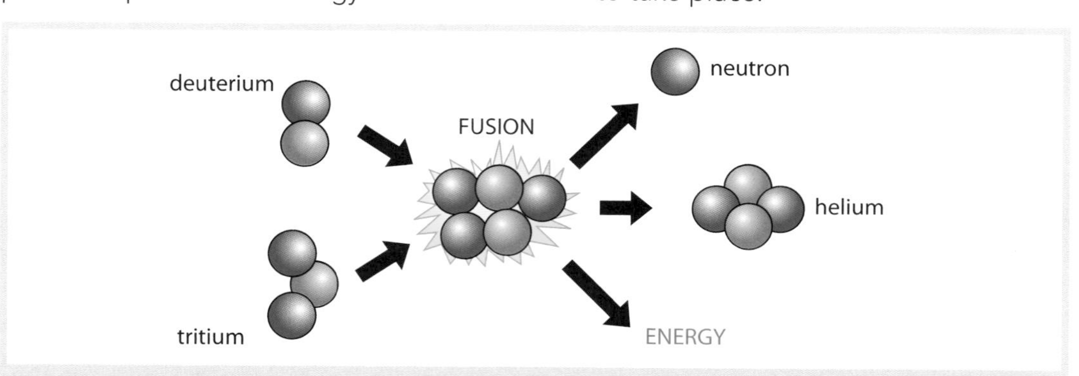

The fusion process. In the Sun, atoms of 'normal' hydrogen (^1H) fuse to form helium. This diagram shows a fusion reaction between two other isotopes of hydrogen called deuterium and tritium. The nuclei of these two isotopes combine to form helium, with the release of a neutron and energy. This is the reaction most often used in fusion reactors on Earth.

Fusion on Earth: the energy source of the future

It is hard to build a fusion reactor on Earth because of the high temperatures and pressures needed. Such high temperatures would melt any container the reaction took place in.

Scientists have managed to achieve controlled fusion in two different ways. In one method, the fusion fuel is heated and pressurised using magnetic fields. The magnetic fields also stop the incredibly hot gases (plasma) from touching the sides of the reaction vessel. In the second method, powerful, crossed lasers are used to achieve fusion on a very tiny scale.

The fuels most often used for fusion reactions are two isotopes of hydrogen, called deuterium (two neutrons) and tritium (three neutrons). These are combined to form helium. The process produces much less radioactive waste than fission. However, so far all fusion reactors use more energy getting the fuel up to the right temperature and pressure than they produce from fusion.

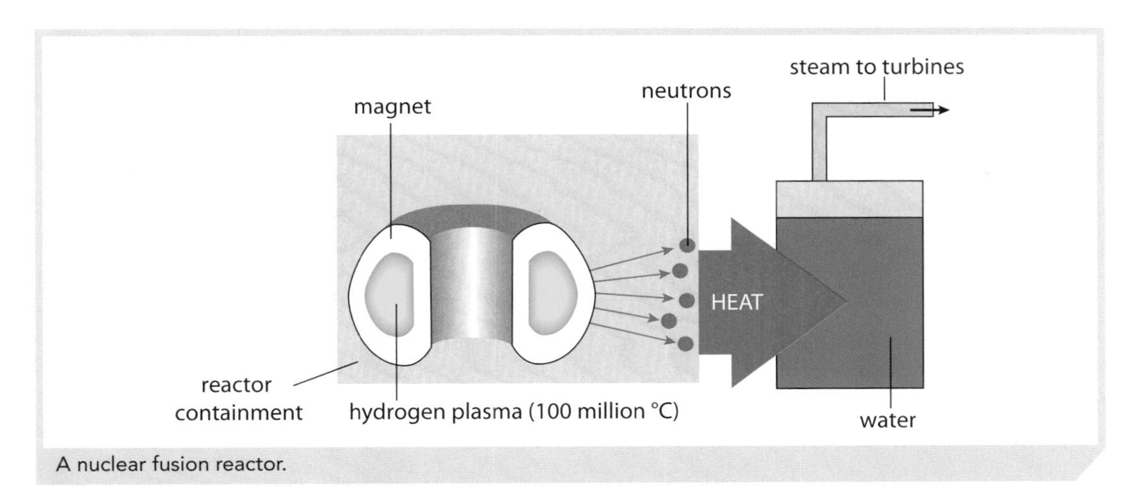

A nuclear fusion reactor.

Activity A

Use the Internet to help you answer the following questions about whether fusion power stations can be cost-effective.

1 When do you think it is likely that there will be a nuclear fusion reactor that can produce energy to power our homes and factories?

2 What are some of the barriers that may prevent nuclear fusion reactors being constructed for commercial use?

Tips

Make sure you link the nuclear equations to show how nuclear fusion takes place.

Task 2 can be attempted using a labelled diagram.

For 2A.P4 you need also to complete task 2 in Assessment activity 3.4.

Assessment activity 3.5

| 2A.P4 (part)

You are a technical assistant working for an international atomic agency. You have been asked to produce a poster that can be presented to members of the public to give them a better understanding of nuclear energy.

1 Use nuclear equations and a labelled diagram to describe nuclear fusion.

2 Explain what happens in a controlled fusion reaction, both in the Sun and in a nuclear fusion reactor.

Lesson outcomes

You should be able to describe nuclear fusion and how controllable nuclear fusion reactions are.

3.7 Investigating electrical circuits

Get started

When you go into a room and switch on the light, what happens? The bulb and the light switch are part of an electrical circuit. Discuss with a partner what other devices at home or in school use electrical circuits.

Key terms

Series – When electrical components are connected so that there is only one route the current can take through the components connected together.

Parallel – When electrical components are connected so that there is more than one route the current can take through the components.

Gradient – Steepness of a line. Resistance is represented by the gradient of a voltage–current graph.

Current – The movement of electrons around an electrical circuit.

Voltage – The amount of energy transferred to an electrical component.

Resistance – How easy or difficult it is for an electric current to flow through something.

What is electricity?

You already know that all materials contain atoms and that atoms are made up of a nucleus surrounded by electrons. Electricity is the movement of electrons through conducting materials (usually metals). The flow of charged electrons around a circuit is called the **current**. The circuit must be complete, with no breaks in the wire, for the current to flow.

Part **(a)** of the diagram below shows an electric circuit with two bulbs, wires and a **cell**. You can compare the amount of energy transferred to different electric components, such as two bulbs, by measuring the **voltage** across each component in turn. This will show you the difference between the amount of energy going into the component, and the amount of energy coming out.

The voltage across different components is not always the same. This is because it is easier for current to flow through some appliances than others. The **resistance** is a measure of how easy it is for current to flow through an electrical component.

Series and parallel circuits

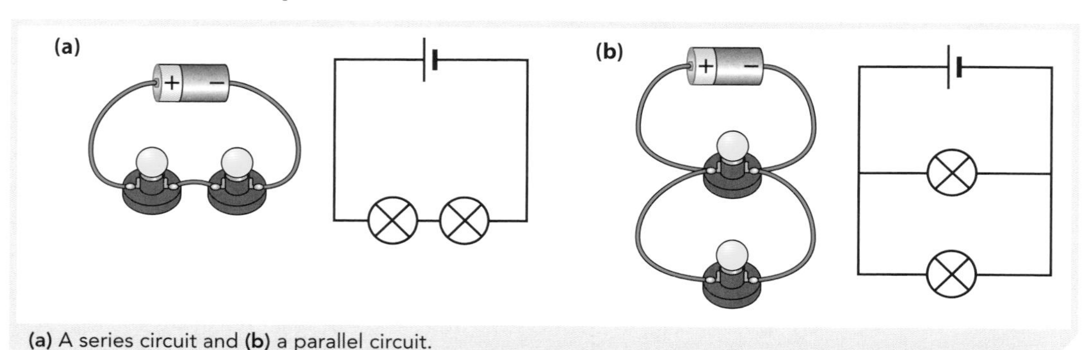

(a) A series circuit and **(b)** a parallel circuit.

Electrical components such as bulbs can be connected next to each other as shown in **(a)** above. They are said to be in **series**. Another way of connecting bulbs is in **parallel**, as shown in **(b)**. The main lights in your home are connected in parallel. Lights on Christmas trees used to be connected in series – they are now more usually in parallel.

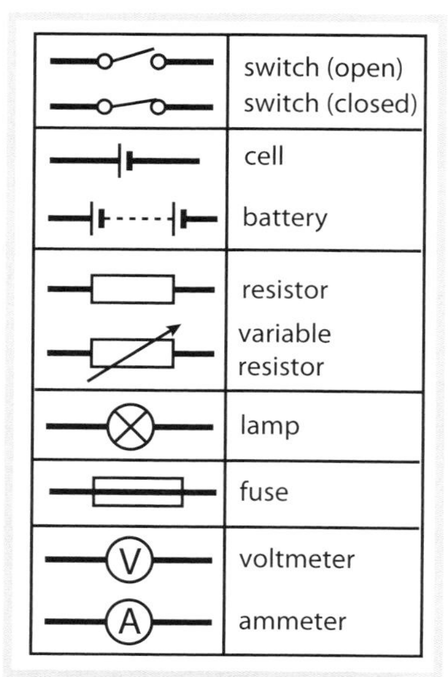

Some common circuit symbols.

If two bulbs are connected in series, they will shine less brightly than a single bulb. This is because the electrical energy that is supplied by the cell is shared between the two bulbs. The additional bulb gives the circuit more resistance so the current will be smaller.

If you add more than one bulb to a parallel circuit the bulbs will be just as bright, but the current flowing through the circuit will be higher. This higher current means the battery will get used up faster than if there was only one bulb.

The source of electrical energy used by the bulbs is the cell. The output of the cell is measured using a **voltmeter**. Voltmeters are connected in parallel.

The current flowing through the bulbs is measured using an **ammeter**. Ammeters are connected in series.

As the voltage in a circuit changes, the current will change. Voltage is measured in volts or millivolts (mV, one thousandth of a volt) while current is measured in amps or milliamps (mA, one thousandth of an amp).

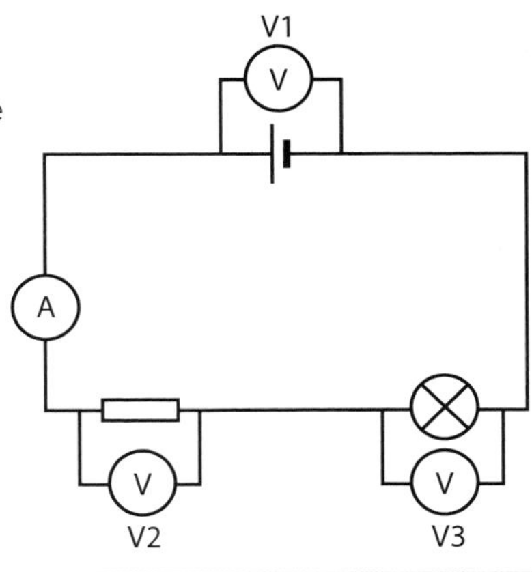

How current and voltages are measured in a series circuit. Voltmeter V1 measures the voltage supplied by the cell. V2 measures the voltage across the resistor, and V3 measures the voltage across the bulb. The current is the same throughout the circuit.

Ohm's law

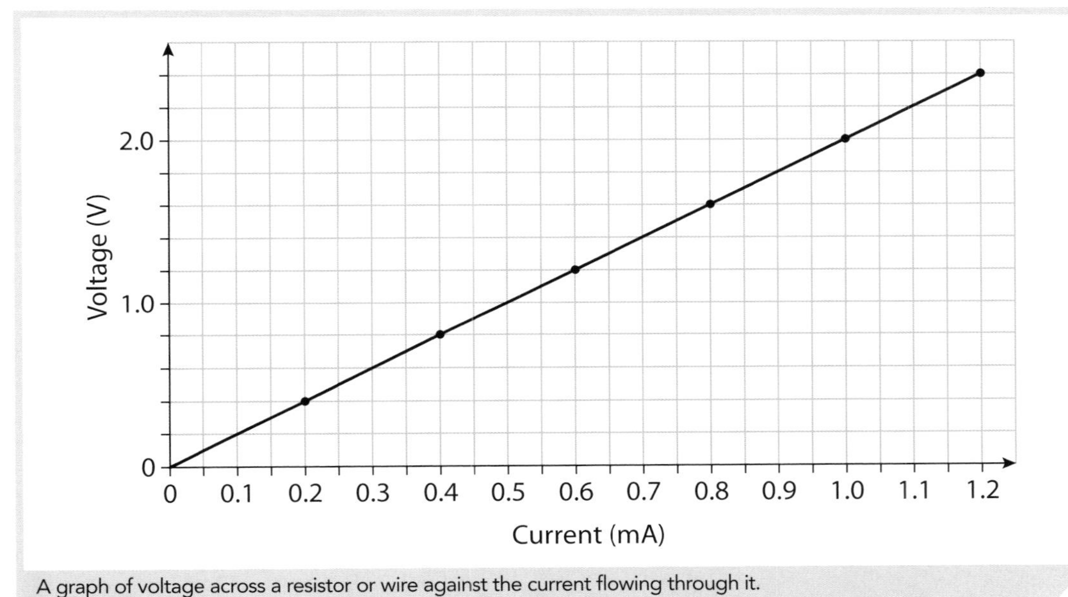

A graph of voltage across a resistor or wire against the current flowing through it.

The graph shows the voltage across a **resistor** plotted against the current through it. The straight line through the origin shows that the current is directly proportional to the voltage. You would get a similar graph using a piece of wire instead of a resistor. This relationship only works if the resistor or wire doesn't get hot.

Ohm's law states that if the resistance is constant, the current is proportional to the voltage. The equation below can be used to represent the relationship between the voltage, current and resistance:

$$V = I \times R$$

where V is the voltage, measured in volts, I is the current in amps, and R is the resistance in ohms (Ω). The **gradient** of the graph represents the resistance, R.

Activity A

What is the voltage across:

(a) an 18 Ω resistor if a current of 2.0 A flows through it?

(b) a 200 Ω resistor carrying a current of 15 mA? (Remember: 1 mA = 1/1000 A.)

Lesson outcomes

You should understand the difference between series and parallel circuits, and be able to predict electrical quantities using Ohm's law.

3.8 Power in an electrical circuit

Link

You learned in lesson 1.24 that power is the amount of energy transferred per unit of time, and that it is measured in watts (W).

This drill has a power rating of 240 W. What is the current in the drill circuit?

Electricity is a form of energy. So electrical power is the amount of electrical energy produced per unit of time. The power in an electrical circuit can be measured by this equation:

$$\text{power (in watts, W)} = \text{voltage (in volts, V)} \times \text{current (in amps, A)}$$

$$P = V \times I$$

Activity A

All electrical devices at home will have electrical information printed on them. Devices such as electric kettles, food mixers, cookers and electric heaters run on mains electricity, which has a voltage of 230 V.
Look at a few of these devices and calculate the following:

(a) the current that is used in normal operation

(b) the resistance.

Note that you will need to use the equations in this lesson!

Worked example 1

Consider the circuit below. What is the current going through the ammeter?

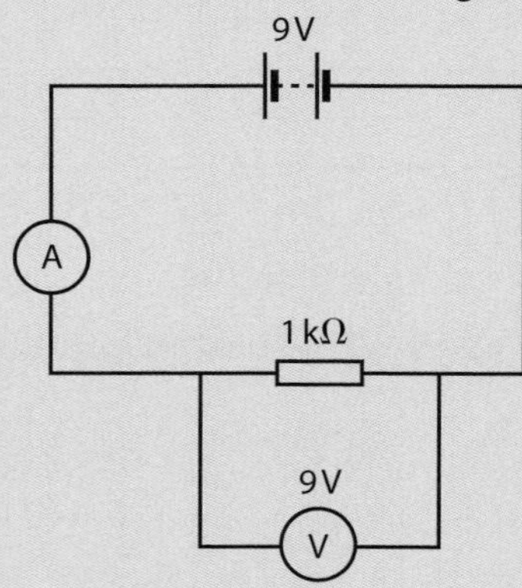

Step 1 The voltmeter is reading 9 V, so the current can be found using Ohm's law:

$$\text{voltage} = \text{current} \times \text{resistance}$$

Step 2 Use algebra to make the current the subject: current = voltage/resistance

Step 3 Substitute the numbers into the equation: remember 1 kΩ is 1000 Ω.

Step 4 Current = 9 V/1000 Ω = 9.0×10^{-3} A or 9 mA.

Worked example 2

Two learners investigate the properties of a wire. They draw a graph of voltage against current, shown below.

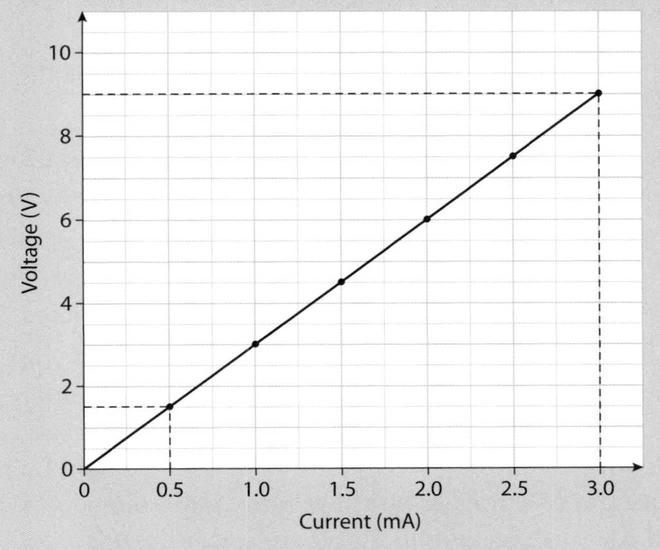

(a) How does the voltage change when the current changes?

The graph is a straight line, so Ohm's law applies. The voltage is proportional to the current as the temperature of the wire is constant.

(b) Use the graph to work out the resistance of the wire. Note that, because this graph is a straight line, we can choose any two voltage measurements from it and use them in our calculation. In this example, we have chosen 9 V and 1.5 V.

Step 1 Resistance is the gradient = change in y-axis/change in x-axis.
Step 2 Resistance = change in voltage/change in current.
Step 3 Substitute numbers: $R = (9\,V - 1.5\,V) / (3\,mA - 0.5\,mA) = 7.5\,V/2.5\,mA = 3\,\Omega$.

(c) Calculate the power when a current of 0.5 A is flowing.

Step 1 From the graph, with a current of 0.5 A, the voltage across the wire is 1.5 V.
Step 2 Power = voltage × current = 1.5 V × 0.5 A = 0.75 W.

Assessment activity 3.6 | 2B.P6

Jake is a technician working for a local electronics company. He wants to design and build an electrical system that will be used in re-designed MP3 players. Part of the design is to predict what current and voltage will flow in a circuit that forms part of the electronic system. His circuit is designed to work with a 9 V battery and to have two resistors, both of 4.7 kΩ. The resistors can be arranged in series or in parallel.

1. Draw a circuit diagram of each circuit (series and parallel). Make sure that your diagram shows the symbols of the instruments that will measure the current and the voltage across the resistors.
2. The series circuit has a total resistance of 9.4 kΩ. Use $V = IR$ to predict the current that will flow.

Tip

For 2B.P6 remember to change units (kΩ) to Ohms before using the equation.

Lesson outcomes

You should know the equation for electrical power, and be able to demonstrate building electrical circuits.

3.9 Batteries

Get started

Have you ever looked at the battery of a mobile phone? Have a look at a battery and read what it says on the back (make sure you turn the phone off first). Have you thought about why it has to be recharged regularly?

Key terms

Electrolyte – The ionic material separating the positive and negative poles inside a battery or cell.

Electrode – The positive or negative pole of a cell or battery.

Direct current (d.c.) – Current that only flows in one direction. It may increase or decrease.

Cell – An electrical cell is a single unit containing two electrodes separated by an electrolyte.

Battery – A battery is two or more electrical cells working together.

Producing electricity

Electricity can be produced by many methods – for example, chemically in batteries, by generators that rotate conducting coils in a magnetic field, and in solar cells that generate electricity from light.

Batteries (non-rechargeable)

In an electric **cell** or a **battery**, two **conductors** are separated by an ionic solution (a solution containing charged atoms or ions). The solution is called an **electrolyte**.

An example is a zinc–carbon cell, where the electrolyte is a damp ammonium chloride paste and the conductors are a carbon rod surrounded by manganese dioxide and the zinc case. The positive and negative conductors are known as poles or **electrodes**.

The diagram shows a zinc–carbon cell connected into a circuit. When the circuit is complete, chemical reactions inside the cell lead to electrons being released at the negative electrode. At the positive electrode electrons are taken up. Electrons flow from the negative electrode, through the wires and the light bulb, and into the positive electrode. The electrons always flow in the same direction, which is known as **direct current (d.c.)**.

These cells are known as non-rechargeable because once the chemicals inside are used up, the cell cannot be re-used.

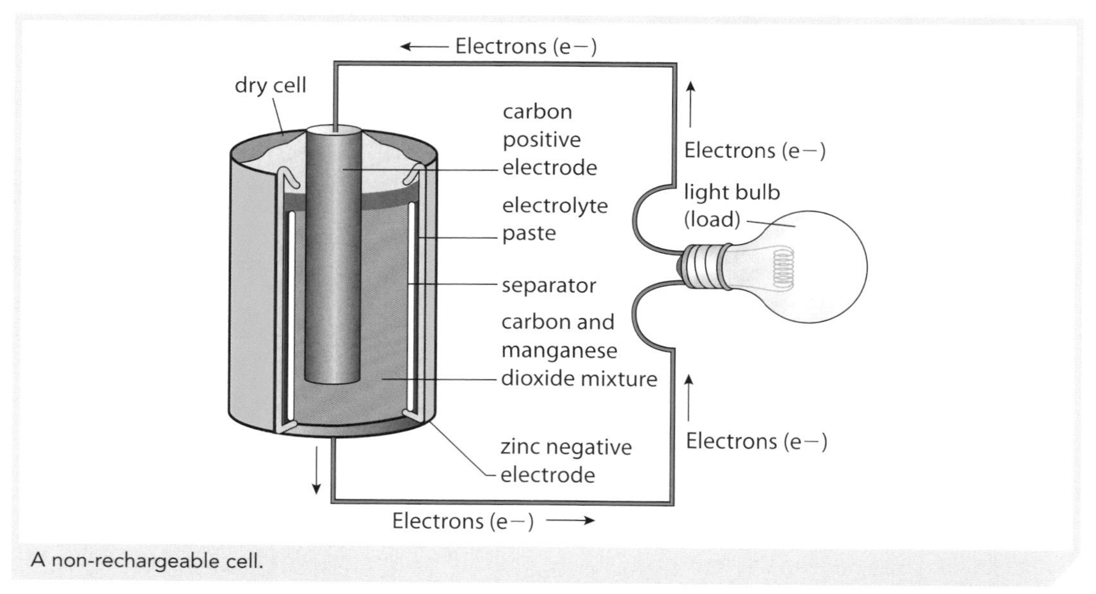

A non-rechargeable cell.

Activity A

Is it possible to get electricity out of a lemon? You can investigate this by inserting two different metal electrodes into a lemon.

1. Design an investigation to measure whether the lemon battery produces a voltage.
2. Check your plan with your supervisor and then carry it out.
3. Record your results. Do the types of metals you use make a difference?
4. Can you make electricity using other fruits?

Rechargeable batteries

Rechargeable cells or batteries work in a similar way to the zinc–carbon cell. However, the chemical reactions that take place in a rechargeable cell are reversible. Electrical energy can be used to reverse the chemical reactions and recharge the cell.

Rechargeable cells and batteries are widely used. Lithium-ion batteries are used in devices such as mobile phones and laptops, because they are light and powerful. They can also be made in a variety of shapes and sizes.

Lead–acid batteries are used in cars and other vehicles. They are heavier than lithium-ion batteries, but they can produce the high currents needed for starting a car. A lead acid battery in a vehicle is continually charging and discharging. Starting the car uses a lot of current, and at night the lights use battery power. However, when the engine is running, some of the engine power is used to recharge the battery.

This electric car has a top speed of 75 mph and a range of about 87 miles. As with other electric cars, the most expensive part is the large lithium-ion battery.

Did you know?

Many modern car makers produce at least one electric vehicle. But there is a problem with electric cars – it takes several hours to recharge the battery.

Researchers at the Massachusetts Institute of Technology (MIT) may have found a solution. They have made experimental batteries that can be recharged in minutes or even seconds. The batteries contain a metal 'sponge' coated with battery materials.

Case study

Ryanne is a junior electronic engineer working for a battery manufacturer. Part of her role is to work in a team developing a new type of longer-lasting battery that can also be recycled. Consumers are asked to send their batteries back to the factory when they are used up. Ryanne selects the batteries that are suitable and begins the process of recycling.

1. Write down two advantages of using this new type of battery compared to a typical battery.
2. What type of electricity would be produced in this new battery?
3. Although it can be recycled, the new battery is not rechargeable. Can you think of a reason why we still use non-rechargeable batteries?

Lesson outcomes

You should be able to explain the differences between non-rechargeable and rechargeable batteries and describe what d.c. is.

Just checking

1. Give an example of **(a)** a type of rechargeable battery and **(b)** where it can be used.
2. Repeat the above question, but for a non-rechargeable battery.

3.10 Other ways of generating electricity

Key terms

Alternating current (a.c.) – Current that flows one way and then the other way. Found in mains electricity.

Photovoltaic cell – a device that produces an electric current when light falls on it.

Efficiency – the useful energy output from a power plant divided by the energy put in.

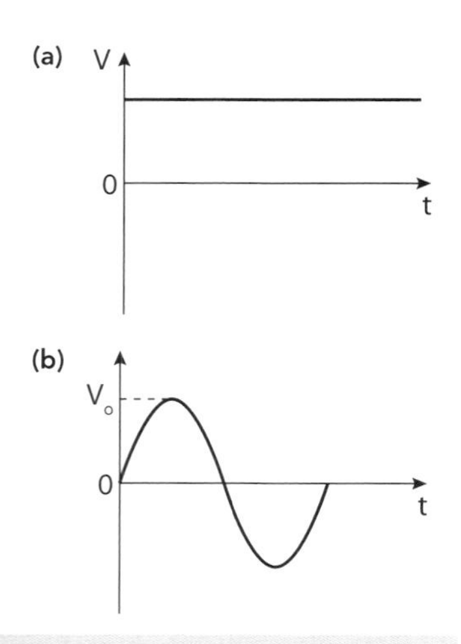

A graph of voltage against time for **(a)** direct current and **(b)** alternating current.

Batteries are only really useful to provide electricity for portable devices. Most of the electrical devices we use run on mains electricity, which comes from power stations. Power stations use energy from various sources to power **generators**.

Generators

If you move a piece of wire in a magnetic field, an electric current flows through the wire. Generators make use of this effect. In a simple generator, a coil of wire rotates between fixed magnets and produces a voltage. If the generator is connected into a circuit, a current flows as the coil turns. After every half-turn of the coil, the current changes direction and flows the opposite way. This type of changing current is called an **alternating current (a.c.)**.

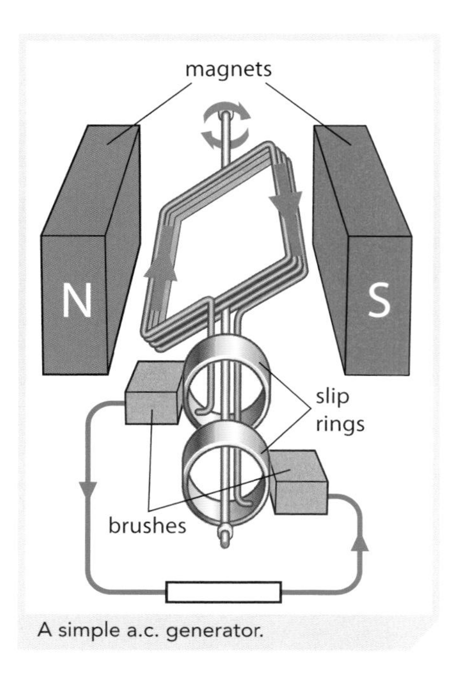

A simple a.c. generator.

A simple generator produces only a small voltage. The voltage a generator produces can be increased by:

- using stronger magnets
- rotating the coils faster
- having more turns of wire on the coil
- having more than one coil
- wrapping the coil around a piece of iron.

The generators in power stations use all these methods to increase the voltage they produce. A large modern generator can produce 25 000 V.

Solar cells

Solar power can be harnessed using **photovoltaic cells**. Photovoltaic cells turn light energy directly into electricity, without the need for a generator. When light falls on silicon crystals within the photovoltaic cell, electrons are emitted. The moving electrons create a flow of electric current.

Some houses and other buildings have arrays of photovoltaic cells fitted to the roof. They produce some or all of the electricity for the building. Solar cells can be used to provide electricity in places where there is no mains supply. They are also used in small portable devices such as solar phone chargers, watches and calculators.

A road warning light powered by solar cells.

Efficiency

Some ways of generating electricity are more efficient than others. **Efficiency** is a measure of how much useful energy is produced compared to the amount of energy put in. So if a solar cell turns half of the light energy falling on it into electricity, the efficiency of the cell is 50%.

Effect on the environment

Any kind of electric power plant has some impact on the environment. Some people think wind turbines are noisy and spoil the view. Building a hydroelectric dam involves flooding land that could otherwise be used for farms and homes. Nuclear power stations produce radioactive wastes that must be stored safely for thousands of years.

However, the biggest environmental impact is from power stations using **fossil fuels**. Coal, oil and gas are all non-renewable sources of energy. Once they run out, no more can be made. Burning fossil fuels emits polluting gases such as carbon dioxide. These gases can cause **acid rain** and contribute to **climate change**.

Worked example

A power station uses gravitational potential energy stored in water from reservoirs in the hills. The average input is 600 MW and the average output is 100 MW. What is its efficiency? Compare this with a typical solar thermal power plant. Remember from lesson 1.24: percentage efficiency is output/input × 100.

Step 1 Write down the equation: efficiency = output energy/input energy

Step 2 Substitute the numbers into the equation:
efficiency = 100 MW/600 MW = 0.167

Step 3 Change to a percentage by multiplying by 100:
efficiency = 100 × 0.167 = 16.7%

Assessment activity 3.7

| 2B.P5 | 2B.M4

1. Identify FOUR methods of producing electricity from different sources.
2. For the methods identified in task 1, describe how they can be used to produce d.c. and a.c. electricity. Use labelled diagrams to help you.
3. Compare the environmental impact and efficiency of using non-renewable energy sources (such as fossil fuels and nuclear) and renewable energy (such as solar technology). Present your findings in the form of a table.
4. A coal-powered station is rated as 1000 MW (output power). Calculate the efficiency of this power station if 3000 MW of power has to be supplied. How does it compare with the efficiency of a typical nuclear power plant?

Just checking

1. How is the electricity generated by an a.c. generator different to that produced by batteries?
2. What kind of electricity do solar cells generate?

Did you know?

Most kinds of electricity generation are less than 50% efficient. Power stations using fossil fuels are only about 40% efficient. A lot of the wasted energy is lost as heat. Nuclear power stations are around 30% efficient, while the best solar cells are only about 6% efficient. Large hydroelectric power stations can be up to 90% efficient.

Link

You have already learnt about renewable and non-renewable sources of energy in lessons 1.25, 1.26 and 2.16.

See lesson 3.5 for more information on accidents in nuclear power stations.

Tips

Task 2 for 2B.P5 is best attempted with a diagram, including showing how voltage changes with time.

For 2B.M4 remember to include the impact that building the power plant has on the countryside, as well as the noise pollution.

Lesson outcomes

You should be able to describe how d.c. and a.c. electricity can be generated from different sources.

3.11 Transmitting electricity

Key term

Transformer – A device that increases (step-up) or decreases (step-down) voltage.

Did you know?

In a step-up transformer the voltage is increased, so you seem to get something for nothing. But remember that:

power = voltage × current

So if a transformer doubles the voltage and the power stays the same, then the current is halved.

The electricity produced in power stations is used around the country. It has to be transmitted from place to place through power lines. All the power stations feed the electricity they produce into a network called the National Grid. This network makes it possible to distribute electricity wherever it is needed.

Minimising energy losses

When electricity flows through a wire, there is some resistance to the flow. This resistance causes heating. Energy that goes into heating the wire is wasted.

Power stations in the National Grid generate a.c. electricity at a voltage of 25 000 V and a frequency of 50 Hz. If electricity is transmitted at this voltage there are large energy losses due to heating in the power cables.

The best way to reduce energy losses due to heating is to reduce the current. This can be done by increasing the voltage. The relationship between power, voltage and current is:

power = voltage × current

Therefore current = power ÷ voltage

If the voltage is doubled and the power remains unchanged, the current is halved.

Transformers

The voltage of a.c. electricity can be changed using a **transformer**. A step-up transformer increases the voltage, while a step-down transformer reduces the voltage.

In the National Grid, step-up transformers increase the voltage from a power station to as much as 400 000 V, to reduce energy losses in long-distance transmission. When the electricity reaches our homes, offices and factories, step-down transformers reduce the voltage.

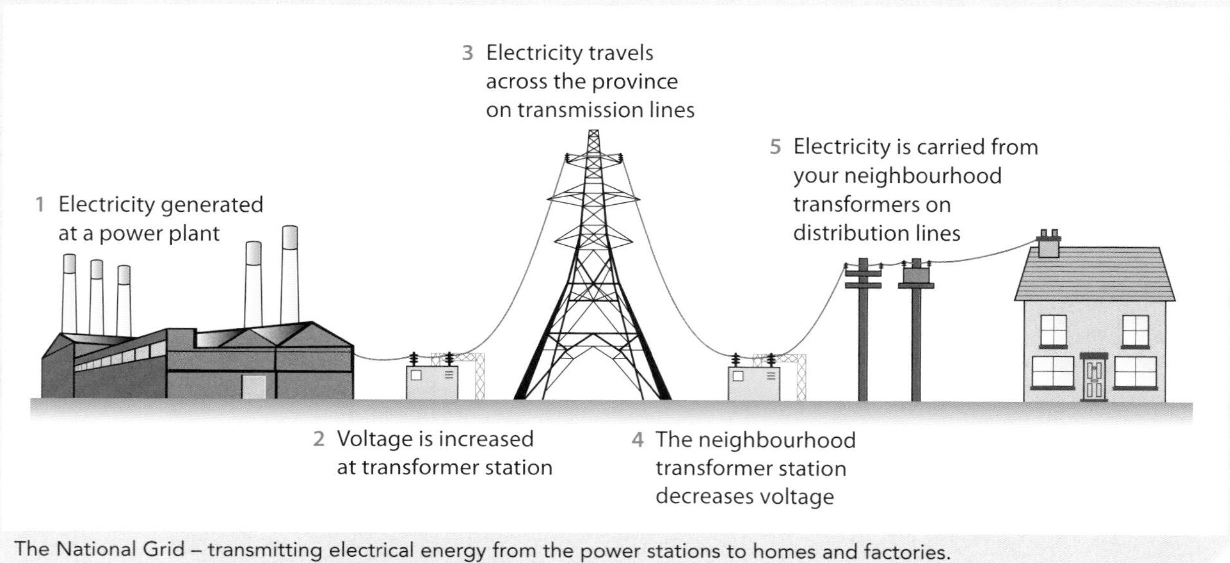

The National Grid – transmitting electrical energy from the power stations to homes and factories.

Activity A

If you look closely around your neighbourhood or near to school, you can locate a transformer that is reducing the electricity supply to 230 V:

- make a sketch of the transformer
- make a note of anything that is written on it.

Also, investigate what kind of transformer is inside a mobile phone recharger power supply.

Case study

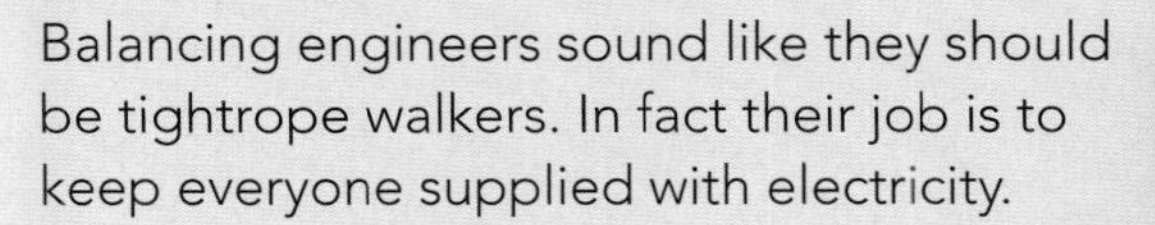

Balancing engineers sound like they should be tightrope walkers. In fact their job is to keep everyone supplied with electricity.

The amount of electricity used across the country varies all the time. More electricity is used in winter than in summer. In the evening electricity demand goes up when millions of people switch on electric lights. Even the TV can affect demand. When a popular programme finishes, millions of people across the country turn on their kettles and there is a sharp rise in electricity use.

Balancing engineers make sure that the supply of electricity balances the demand.

Balancing engineers must constantly track how electricity demand is changing. When more electricity is needed they can call on hydroelectric or gas turbine power stations, which can increase electricity production within minutes.

Assessment activity 3.8

| 2B.P7 | 2B.M5 | 2B.D4

You are an assistant technician working for the technology laboratories of the National Grid. You have been asked to explain the main concepts of electrical transmission. You need to cover the following points in your explanation.

1. Describe electrical power, in terms of transmitting electricity.
2. Using labelled diagrams, describe how electrical energy is transferred to the home or industry.
3. Discuss how energy loss can be minimised during the transmission of electricity to the home and workplace.
4. Assess what you have discussed in task 3, using equations and numerical values.

Just checking

1. Explain why electricity is transmitted at high voltages.
2. Explain why using thin copper cables would not be an efficient way of minimising energy loss.

Tips

For 2B.P7 don't forget to include not only step-up transformers, at the power plant, but also step-down transformers close to homes and factories.

Tasks 3 and 4 for 2B.M5 and 2B.D4 can be attempted together and should include the voltage used for transmission and the thickness of cables. The important point for 2B.D4 is that equations are used, as mentioned in this section. Don't forget to explain the symbols used.

Lesson outcomes

You should know how energy can be transmitted to homes and the work place, and how loss is minimised.

3.12 A journey into our Solar System

A solar system consists of a star and the planets and other bodies that go around it. The star at the centre of our Solar System is the Sun. It makes up nearly 99% of the Solar System's mass. The Sun's gravity keeps the planets in their orbits.

Our Solar System. The planets are shown to scale, but their orbits are much further apart. Mercury is nearest the Sun, followed by Venus, Earth, Mars, Jupiter, Saturn, Uranus and Neptune.

Composition of our Solar System

Planets

The eight planets orbit the Sun in elliptical orbits. The time for one orbit is the planet's year, and the time for one rotation on its axis is the planet's day. If you visited Saturn, for example, a day trip would last only 10 Earth hours, but a year's visit would last almost 30 Earth years.

Natural satellites

Natural satellites are moons or other objects that orbit a planet. Jupiter has about 70 moons, but more are being discovered all the time.

Dwarf planets

Pluto, Ceres and Eris are dwarf planets. Ceres is in the asteroid belt between the orbits of Jupiter and Mars. Eris is about the same size as Pluto and, like Pluto, orbits beyond Neptune.

Asteroids

These rocky objects, smaller than a planet, are mostly found in the asteroid belt. Fragments of **asteroids** that burn up in the Earth's atmosphere are known as **meteors**. Those that are big enough to survive until they hit the Earth are called meteorites. It is thought a large meteorite hit the Earth about 65 million years ago, causing climate change and destroying 75% of living things, including the dinosaurs.

Comets

These are balls of ice and dust that come from beyond Pluto and travel in very elliptical orbits. When **comets** are close to the Sun the ice evaporates and forms a 'tail' that can sometimes be seen with the naked eye.

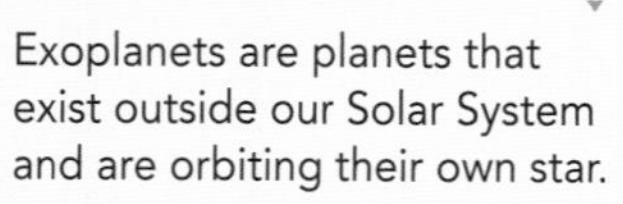

Exoplanets are planets that exist outside our Solar System and are orbiting their own star.

1 Find the names of some of the planets that have been discovered recently, and investigate how they compare to the planets in our Solar System.

2 Present your results in the form of a poster, including how far the planet is from its star, how long it takes to orbit its star and whether it is a rock or gas planet.

How our Solar System was formed

We cannot be sure how the Solar System formed, but scientists have proposed different models to explain how it happened. One model is the widely accepted nebular theory of planetary formation.

- About 4500 million years ago, a huge cloud of dust, ice and gas (a nebula) was pulled together by gravity.
- At a certain point, gravity caused the cloud to collapse inwards, becoming about 1000 times smaller.
- As the nebula got smaller and more dense, the core became hotter and the cloud formed a rotating disc (a protostar).
- Eventually the temperature and pressure at the core was so high that nuclear fusion reactions began. The Sun was formed.
- Some gas and dust did not become part of the Sun, but formed a rotating disc around it.
- The heavier dust in the disc formed the rocky inner planets (Mercury, Venus, Earth and Mars).
- The lighter dust and gases came together to make the giant gaseous outer planets (Jupiter, Saturn, Uranus and Neptune). Scientists are unsure of the exact timing, but the whole process took millions of years.

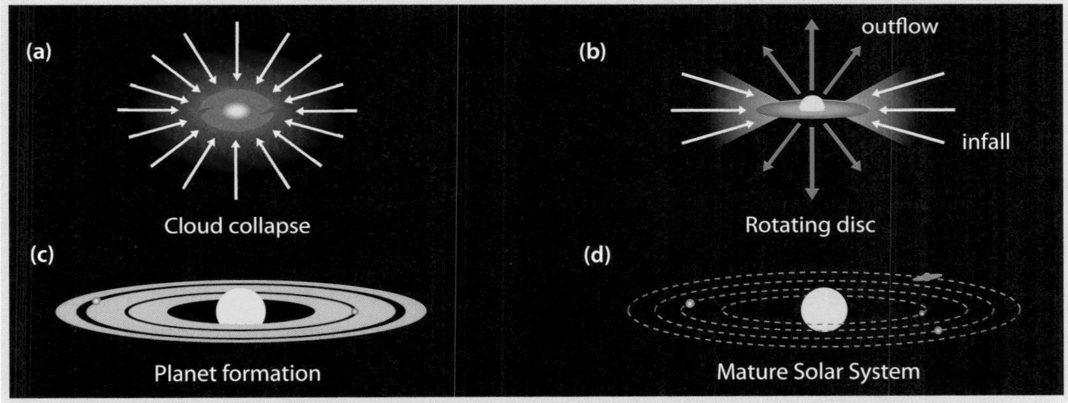

The nebular theory of planetary formation.

Did you know?

The spacecraft Kepler was launched in 2009 with a mission to survey part of the sky looking for planets orbiting other stars. Kepler identifies possible exoplanets by looking for the tiny dip in brightness that happens when a planet crosses the face of a star. It has identified over 2000 possible exoplanets.

Assessment activity 3.9

| 2C.P8 (part)

You are working in the National Space Centre and you are in charge of public relations. As part of your role, introducing members of the public to space, carry out the following tasks.

1 Produce a diagram or model of our Solar System. Include the order of the planets and describe how your model would be different if it were to scale.

2 Explain the differences between an asteroid and a comet. Using your model, describe the structure of the Solar System, including the relative distances and sizes of the objects.

Tip

Don't forget to include asteroids, comets and dwarf planets when describing the Solar System.

Just checking

1 Which planet's year is nearly 30 Earth years?
2 What is the nebular theory of planetary formation?

Lesson outcome

You should be able to identify the components of our Solar System.

3.13 Methods of observing the Universe

Key term

Light year – The distance that light travels in a year – about 9.5 thousand billion km or about 6 thousand billion miles.

Link

This lesson links to lessons 3.1 and 1.28

When we look at the night sky we are looking back in time. Light arriving at our telescopes on Earth may have taken millions or even billions of years to reach us. What we see is what the star looked like many years ago.

Stars and galaxies are many **light years** away so it is impossible for us to go and investigate them ourselves. However, the light they produce can be observed and analysed on Earth.

Satellites and robotic probes can be sent into orbit and can even land on other planets in the Solar System. They can use cameras and other sensors to send back information about the planet's structure and atmosphere.

Most telescopes are based on Earth, but there are a number of telescopes in space. Telescopes such as the Hubble Space Telescope, which is 570 km (or 354 miles) above the Earth's surface, have given us amazing pictures of stars and distant galaxies.

Activity A

The Lovell Telescope is a famous radio telescope that is located at Jodrell Bank near Manchester. It is the third largest steerable telescope in the world.

1. Using the Internet to help you, draw ray diagrams to show how the Lovell Telescope forms an image.
2. Write down two of the important discoveries that have been made using the Lovell Telescope.
3. Explain why the Lovell Telescope can be used both in the day and at night.

Using electromagnetic radiation

Stars emit not just visible light, but all forms of electromagnetic (EM) radiation, such as radio waves, infrared and X-rays. The table on page 149 describes different types of telescopes and space probes, and the types of EM radiation that they detect.

Optical telescopes are relatively cheap compared to other types, but they are only useful at night and in clear skies. Radio telescopes have the advantage that they can be used both in the day and night, and radio waves are not blocked by clouds or pollution.

Infrared telescopes can detect the heat of newly formed stars in gas clouds that a light telescope cannot see through. X-ray and gamma ray telescopes can detect super-hot objects such as the remains of an exploded star or the gases swirling into a black hole.

The Hubble Space Telescope.

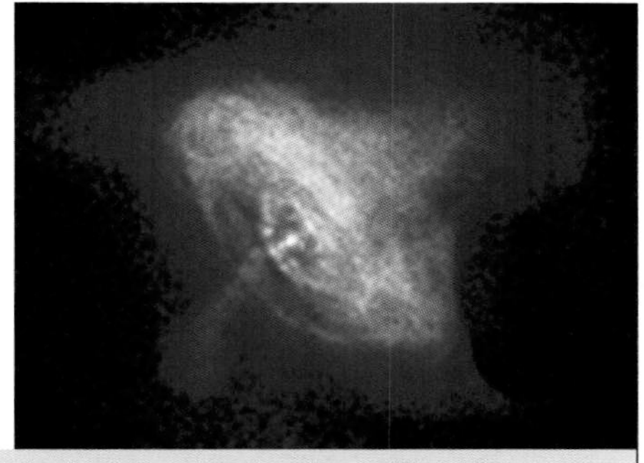

The Crab Nebula is the remains of an exploded star. The picture on the left was taken by the Hubble Space telescope. It shows the gas cloud left behind after the explosion. The photo on the right was taken with the Chandra X-ray telescope. It shows a tiny star remnant at the centre of the gas cloud, sending out large amounts of X-ray radiation.

Type of EM radiation detected	Location	Comment on operation	An example
Optical	Ground	Reflecting: uses concave mirrors to reflect light to a detector	GTC (Gran Telescopio Canarias). It is the biggest in the world, with a diameter of about 10 m. It is located in La Palma, Canary Islands.
Optical	Space	Can also detect light in near IR and UV regions	Hubble Space Telescope
Radio (reflecting)	Ground	The reflectors are large concave metal dishes. Often many small radiotelescopes are used together.	VLA (Very Large Array) observatory in New Mexico (USA): dishes are connected electronically in networks to give an equivalent diameter of about 30 km.
Infrared (IR)	Space	Infrared radiation is absorbed by the Earth's atmosphere so telescope must be in space. Expensive because if a fault develops it can only be fixed by astronauts.	IRAS (Infrared Astronomical Satellite)
X-ray	Ground (high attitude)	Cloud can partially block X-ray light, so needs to be at high altitude on Earth or flown in balloons.	
X-ray	Space		Chandra X-ray observatory: looking at black holes and supernovae.
Gamma ray	Space	Gamma ray radiation is absorbed by the Earth's atmosphere so telescopes must be in space. Expensive because if a fault develops it can only be fixed by astronauts.	Swift satellite: It looks at explosions occurring in the Universe and the gamma ray bursts they produce. It can also detect light in X-ray, UV and visible regions.

Assessment activity 3.10 | 2C.P9

You have a job as an assistant technician in NASA's space centre. They are looking at designing a new telescope and would like you to investigate what telescopes are available at the moment. Your tasks are as follows.

1 List *three* types of telescope that can be used to observe the Universe. These need to include some located on Earth and some in space.
2 Describe how these three telescopes are suitable for their use.

Tip

For 2C.P9 just describing the telescopes is not enough – you need to explain what makes them suitable.

Lesson outcome

You should be able to identify and describe methods of observing the Universe.

3.14 Our Galaxy as part of the Universe

Key terms

Galaxy – Billions of stars that are held together by the force of gravity. They tend to move about a central mass.

Did you know?

If we could travel at the speed of light, it would take about 1 second to travel to the Moon, 8 minutes to the Sun, but over 4 years to get to our next nearest star!

Remember

1 billion is equal to 1 thousand million or 1 000 000 000 or 10^9.

Our Sun is just one of billions of stars in our **galaxy**, which is called the **Milky Way**. In fact, astronomers believe there are about 100 billion stars in the Milky Way.

Our nearest star, after the Sun, is called Proxima Centauri and is 4.2 **light years** away. That is about 25 thousand billion miles from Earth.

Astronomers believe there must be a lot more material that we can't see yet. This is called dark matter.

Our galaxy is believed to have a spiral shape. We actually live on the edge of one of the spiral arms, about half way from the centre of the galaxy.

How the Milky Way looks through a telescope from Earth. The bright spot in this photo is Halley's comet.

There is some evidence that there is a giant black hole active at the centre of our galaxy. A black hole is a star that has reached the end of its life and is so heavy that it has collapsed into itself in such a way that nothing escapes from it, not even light.

The Universe

Some galaxies have **elliptical** shapes and don't rotate. The biggest galaxies are elliptical. Other galaxies are **irregular** – they don't have a clear shape. The Milky Way belongs to a group of 30 galaxies called the Local Group. Our nearest neighbour, the Andromeda Galaxy, belongs to this group. Another group of galaxies called the Virgo Cluster contains 2500 galaxies.

Andromeda – our nearest galaxy neighbour. It is about 2.5 million light years away and, like our own galaxy, has a spiral structure.

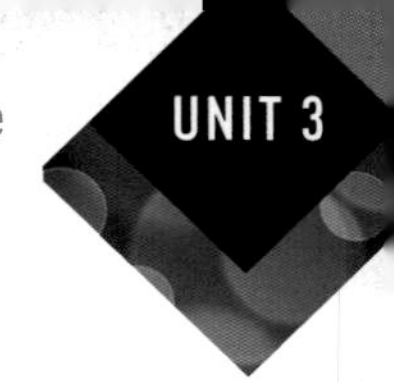

Within our Universe there are billions of galaxies. The Universe has a sponge-like structure in which there are 'filaments' of millions of galaxies. Here is a computer simulation of what this might look like.

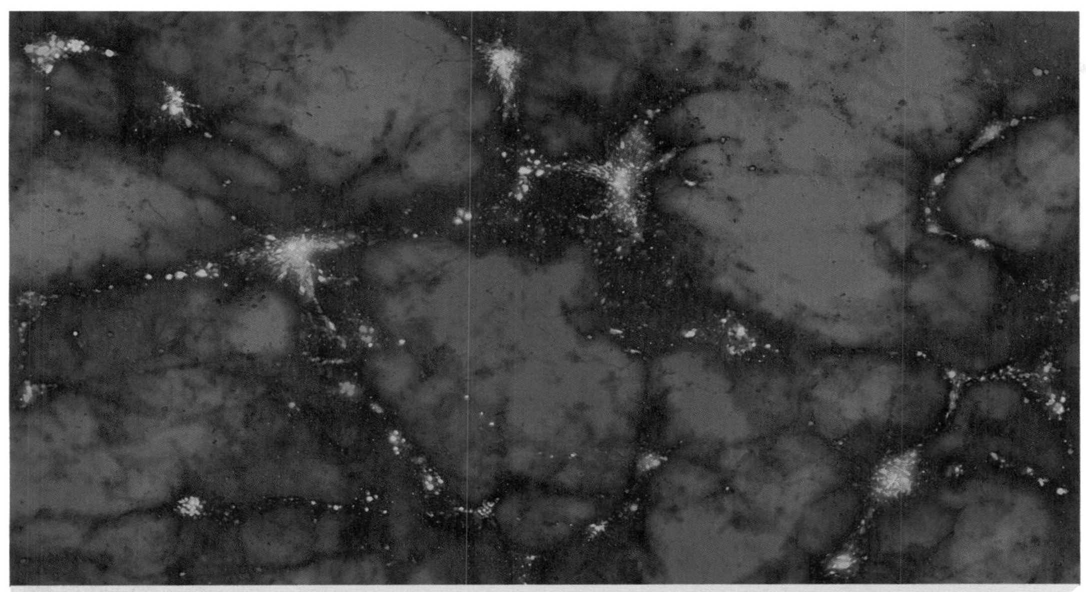

A simulation of the structure of the Universe. The bright knots represent galaxy super clusters.

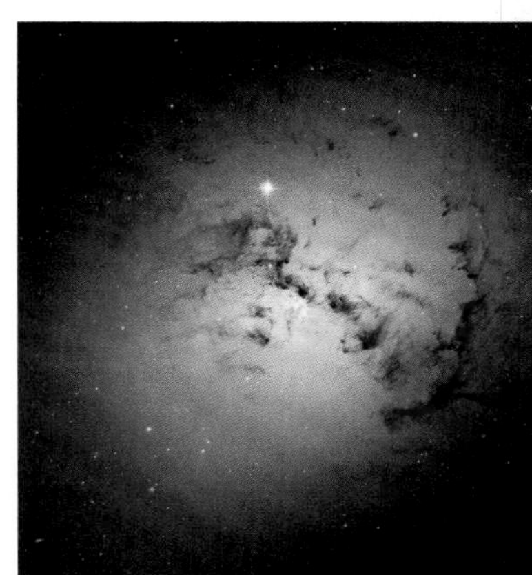

The Dusty Galaxy has an elliptical shape.

Activity A

Using the Internet and scientific magazines, investigate the Whirlpool Galaxy, known as M51. Present your results in the form of a leaflet.

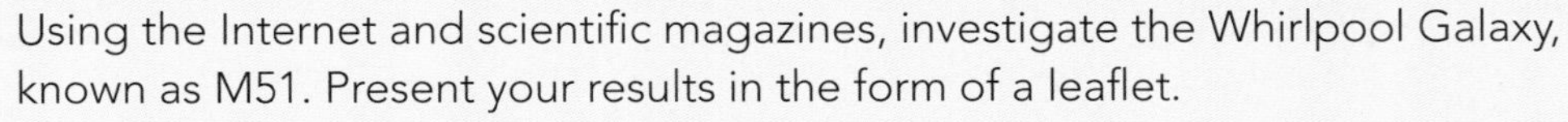

- How far is it from our Milky Way Galaxy?
- Note any similarities between the Whirlpool Galaxy and the Milky Way.
- Where is it located in the night sky?

Assessment activity 3.11

| 2C.P8 (part)

Make a poster to describe the structure of our Universe. Make sure you include the following.

1 A labelled diagram of our galaxy and neighbouring galaxies.

2 A labelled diagram of a model of our Universe.

Tips

For 2C.P8 you need to include a description of not only spiral galaxies, like our own, but elliptical galaxies.

Just checking

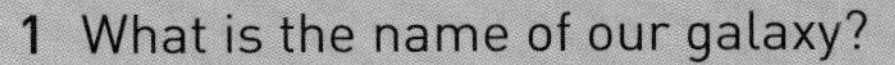

1 What is the name of our galaxy?

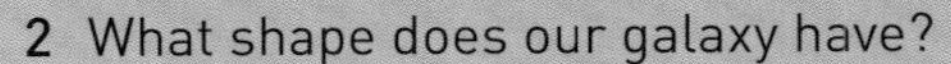

2 What shape does our galaxy have?

Lesson outcome

You should be able to describe the structure of the Universe and our Solar System.

3.15 The changing Universe

Key term

Red shift – The increase in wavelength (decrease in frequency) of electromagnetic radiation from distant, receding galaxies due to the expansion of the Universe.

When you look at the sky at night, you could be deceived into thinking that not much is going on in the Universe. In fact, the Universe is in a state of continuous change, with energetic explosions taking place all the time. Stars have extremely active lives. As they get older they change into red giants, white dwarfs or black holes. Their final destination depends on how massive they were at the beginning of their lives. Our Sun is a smallish yellow star. It is thought to be half-way through its life-cycle.

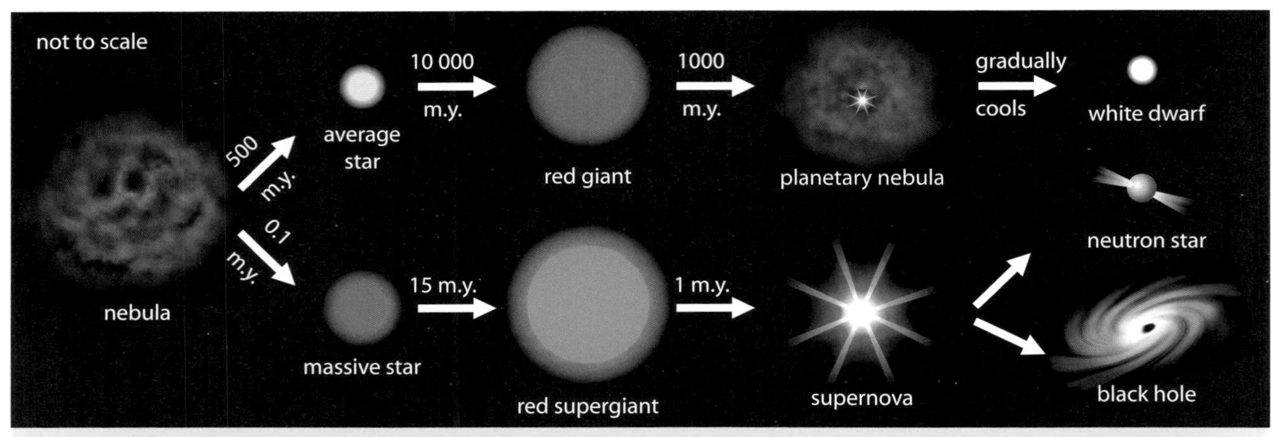

Stars evolve from one type to another. After starting life as a nebula, they begin on the main sequence and then, depending on their mass, they either end up as a white dwarf, a neutron star or a black hole.

Did you know?

Our Sun travels at 220 km/s, which is 500 000 miles per hour.

Red shift

In the spectrum from any star, there are thin black lines called absorption bands. In the 1920s, astronomers noticed differences between these absorption bands in the spectrum of our Sun and spectra from distant galaxies. In spectra from distant galaxies, the black lines were shifted to longer wavelengths. Astronomers called this **red shift**, because the bands were shifted towards the red end of the spectrum.

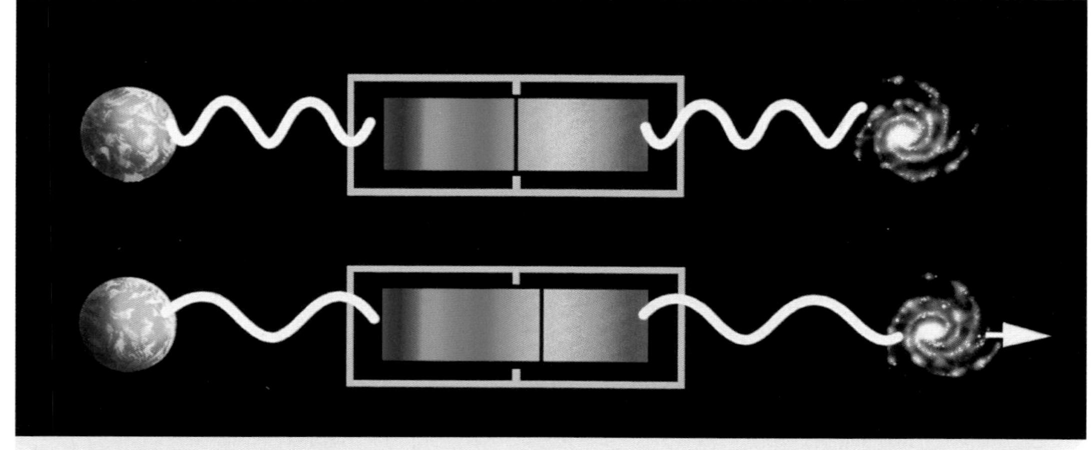

The red shift.

The best explanation for why the light from distant galaxies are red-shifted is that these galaxies are moving away from us. More distant galaxies are red-shifted more than those that are closer.

Space is expanding

The fact that most galaxies are moving away from each other is evidence that the Universe is expanding. However, the galaxies are not moving through space. It is *space itself* that is expanding.

You can get a sense of how space is expanding by drawing some dots on a balloon. Each dot is a galaxy. As you blow up the balloon, the dots move apart. But the dots don't move on the balloon's surface. The space between the dots gets bigger because the surface of the balloon expands. In a similar way, galaxies move apart because space expands.

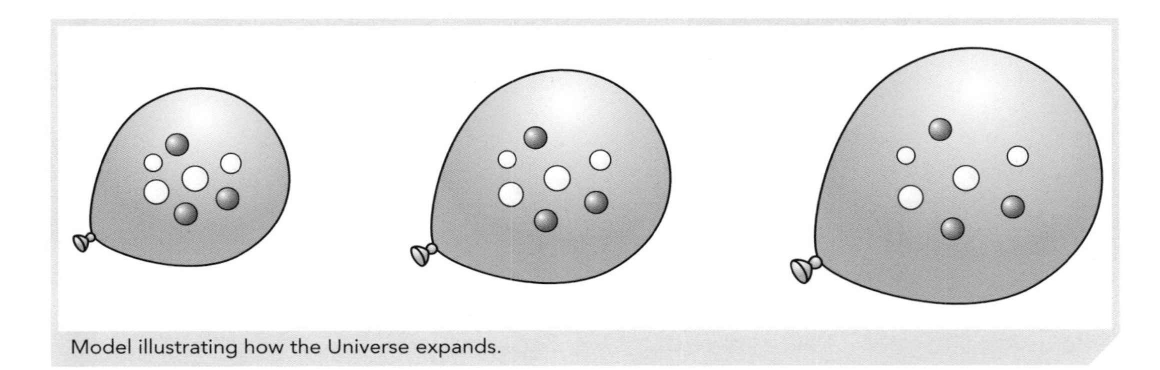

Model illustrating how the Universe expands.

Activity A

As mentioned in lesson 3.14, Andromeda is our nearest neighbouring galaxy. The spectrum for Andromeda shows not a red shift but a blue shift.

1. How do the wavelengths of light coming from Andromeda compare to the wavelengths from a galaxy that is showing a red shift?
2. What do you think the blue shift is telling us about the motion of Andromeda?

Assessment activity 3.12

| 2C.P10 | 2C.M7

1. Describe the dynamic nature of our Solar System and the Universe. You need to refer to the following in your answer:
   - planetary motion
   - stars evolving
   - the Universe expanding.
2. Explain why the evidence shows that the Universe is changing.

Tips

2C.P10 can be achieved by identifying evidence which shows that the universe is changing.

2C.M7 can be achieved using the evidence of CMBR (see lesson 3.16) and the expanding Universe.

Lesson outcome

You should be able to describe the dynamic nature of our Solar System and the Universe.

3.16 The origin of the Universe

Key terms

Cosmic microwave background radiation (CMBR) – Electromagnetic energy that comes from all directions in space, and is believed to have come from the Big Bang.

Big Bang theory – The theory that the Universe began with an explosion.

There are many models that attempt to explain the origin of the Universe. The most successful theory is the **Big Bang theory**, because of the amount of experimental evidence that supports its predictions. According to the Big Bang theory, our Universe began as an infinitely small, very dense, very hot point (called a singularity) which initially expanded incredibly fast.

In the very earliest moments the Universe was too hot for matter to exist. However, as it expanded, the Universe cooled, and particles of matter began to form. The matter eventually came together to form stars and galaxies.

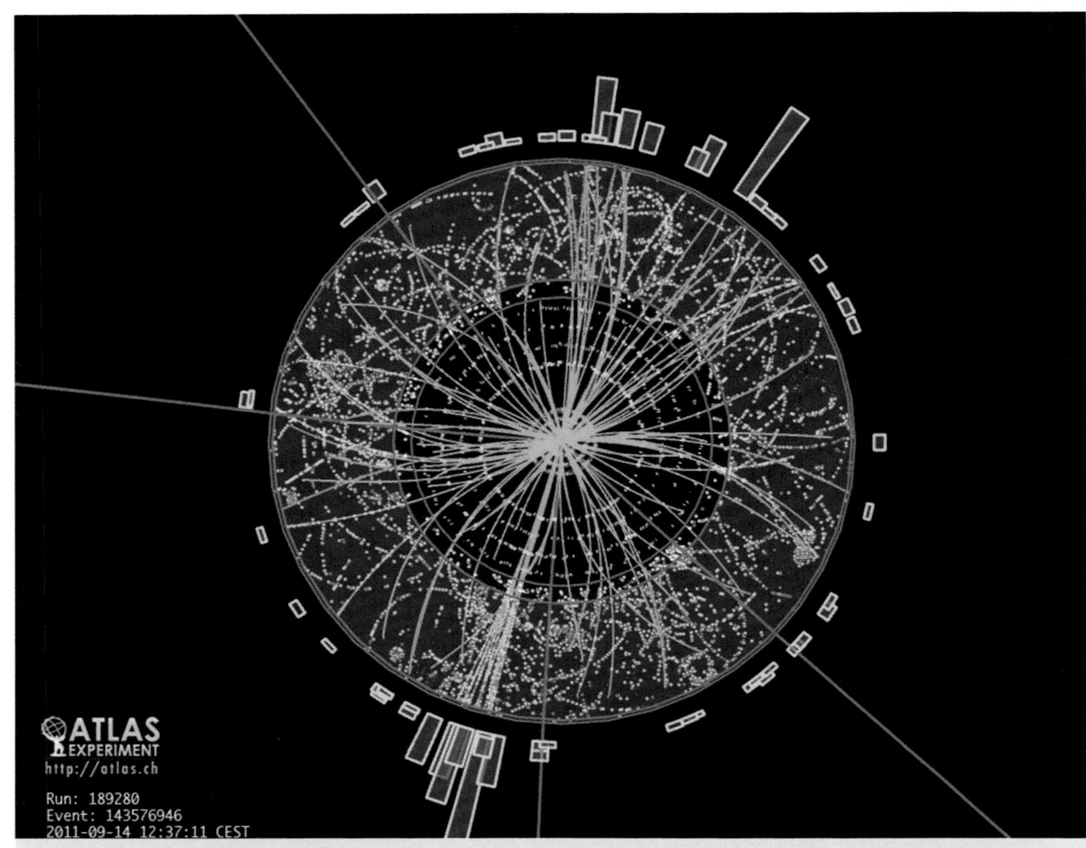

The Large Hadron Collider (LHC) is a huge **particle accelerator** buried deep underground in Switzerland. Inside the accelerator, subatomic particles are smashed into each other at very high speeds. This view is a zoom into the central part of the detector, and shows the results after a collision between two protons. The aim of these experiments is to simulate the conditions shortly after the Big Bang.

Activity A

Do some research on the Internet and in books to find out more about the Big Bang.

1 Find out when the Big Bang is thought to have happened.

2 What is the Big Crunch? See if you can find out.

Evidence for the Big Bang

- As discussed in lesson 3.15, light arriving from distant galaxies is shifted towards the red end of the electromagnetic spectrum. The further away the galaxies are, the larger the red shift. Red shift is evidence that the Universe is expanding. This evidence supports the Big Bang theory.
- If the Universe was initially very hot, it should be possible to find evidence of this burst of high energy. Scientists predicted in the 1940s that this evidence would be a low level of microwave radiation coming from all parts of the sky. This microwave radiation was detected in 1964. It was called the **cosmic microwave background radiation (CMBR)**. The CMBR is strong evidence in favour of the Big Bang theory.

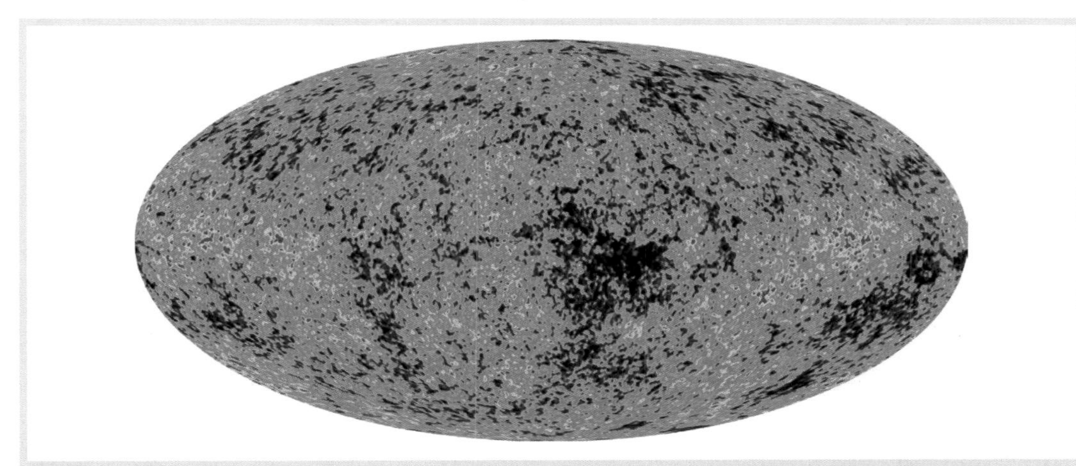

The image shows a detailed map of the CMBR across the whole of space. The different colours represent very tiny differences in the temperature of the CMBR.

- The Big Bang model predicts that there should be an abundance of helium and hydrogen in stars and galaxies, and this has been confirmed through many observations. In fact the Big Bang model predicts that the amount of helium in early stars should be no more than 30%. This also agrees with observations of very old stars.
- By looking at how stars have evolved, from their temperature and chemical make up, the age of stars can be estimated. The oldest stars are about 13 billion years old.

Assessment activity 3.13

| 2C.M6 | 2C.D5

1. Evaluate the evidence for the Big Bang theory.
2. Explain models of how our Solar System and the Universe may have been formed.
   - Use the theory of planetary formation to describe how the Solar System was formed.
   - Use the Big Bang theory to describe how the Universe was formed.

Tip

2C.M6 requires a basic description, without any reference to evidence. The planetary theory is best explained by use of a diagram. You could include Hubble Telescope images showing planets being formed, as described by the planetary theory.

The main point of 2C.D5 is to explore the ideas and evidence that indicate that the Universe was formed according to the Big Bang – you should come up with an evaluation of this evidence, not a description of the Big Bang.

Just checking

1. Explain how galaxies moving away from each other is evidence for the Big Bang theory.
2. Explain the significance of the CMB radiation.

Lesson outcome

You should be able to describe how the Universe and our Solar System were formed.

Introduction

In this unit you will examine what causes variation within species and between species. You will see how all the species living today have evolved from other species. You will study Charles Darwin's theory of natural selection as a mechanism for evolution, and learn how scientists classify living organisms and construct keys for identifying organisms. You will also learn how living things interact with each other and with their environment.

You will find out about some of the ways that humans affect the environment, for example, by clearing land for farming to produce food for us all. You will also look at ways of monitoring and counteracting pollution.

Finally, you will look at some of the factors that affect your health, from infectious and inherited diseases to your lifestyle and environment.

The content of this unit is useful for anyone considering a career in environmental biology, health promotion, sports science or as a personal trainer.

Assessment: You will be assessed using a series of internally assessed assignments.

Learning aims

After completing this unit you should:

a have investigated the relationships that different organisms have with each other and with their environment

b have demonstrated an understanding of the effects of human activity on the environment and how these effects can be measured

c have explored the factors that affect human health.

I really enjoyed this unit, especially the outdoor surveys we did to see how polluted our local stream is. We had to identify organisms in the stream, and I now understand how to use and make keys to identify organisms.

Lucy, *14 years old*

4 Biology and Our Environment

Assessment Zone

This table shows you what you must do to achieve a Level 1 Pass, or a Level 2 Pass, Merit or Distinction grade, and where you can find activities in this book to help you.

Assessment criteria			
To achieve a Level 1 Pass grade, the evidence must show that you are able to:	To achieve a Level 2 Pass grade, the evidence must show that you are able to:	To achieve a Level 2 Merit grade, the evidence must show that you are able to:	To achieve a Level 2 Distinction grade, the evidence must show that you are able to:
Learning aim A: Investigate the relationships that different organisms have with each other and with their environment			
1A.1 Distinguish between variation due to genes and variation due to environmental factors. Assessment activity 4.1	**2A.P1** Describe the role of genes and the environment in variation. Assessment activity 4.1	**2A.M1** Explain the role of genes and the environment in evolution. Assessment activity 4.1	**2A.D1** 2A.D1 Evaluate the impact of genes and the environment on the survival or extinction of organisms. Assessment activity 4.1
1A.2 Construct simple keys to classify organisms. Assessment activity 4.3	**2A.P2** Describe how characteristics are used to classify organisms. Assessment activity 4.3	**2A.M2** Discuss the factors that affect the relationship between different organisms. Assessment activity 4.2	
1A.3 Construct food chains and food webs. Assessment activity 4.2	**2A.P3** Describe the different ways in which organisms show interdependence. Assessment activity 4.2		
Learning aim B: Demonstrate an understanding of the effects of human activity on the environment and how these effects can be measured			
1B.4 Identify human activities that affect an ecosystem. Assessment activity 4.4	**2B.P4** Describe the impact that different human activities have on ecosystems. Assessment activity 4.4	**2B.M3** Analyse the effects of pollutants on ecosystems. Assessment activity 4.4	**2B.D2** Explain the long-term effects of pollutants on living organisms and ecosystems. Assessment activity 4.4
1B.5 Identify living and non-living indicators and the type of pollution they measure. Assessment activity 4.5	**2B.P5** Describe how living and non-living indicators can be used to measure levels of pollutants. Assessment activity 4.5	**2B.M4** Discuss the advantages and disadvantages of methods used to reduce the impact of human activity on ecosystems. Assessment activity 4.6	**2B.D3** Evaluate the success of methods to reduce the impact of human activity on an ecosystem, for a given scenario. Assessment activity 4.6
1B.6 Describe how recycling and reusing materials can reduce the impact that human activities have on an ecosystem. Assessment activity 4.6	**2B.P6** Describe the different methods used to help reduce the impact of human activities on ecosystems. Assessment activity 4.6		

Assessment criteria			
Learning aim C: Explore the factors that affect human health			
1C.7 List the different biological, social and inherited factors that affect human health. Assessment activity 4.9	**2C.P7** Describe how pathogens affect human health. Assessment activity 4.7	**2C.M5** Explain how bacteria can become resistant to antibiotics. Assessment activity 4.7	**2C.D4** Evaluate the use of antibiotics, pedigree analysis and vaccination programmes in the treatment and prevention of childhood illnesses. Assessment activity 4.9
1C.8 Identify measures that can be taken to prevent and treat infectious diseases. Assessment activity 4.7	**2C.P8** Describe two different treatment regimes: one used to prevent a disease and one used to treat a disease. Assessment activity 4.9	**2C.M6** Explain the use of pedigree analysis. Assessment activity 4.9	
1C.9 List some benefits of exercise on health. Assessment activity 4.8	**2C.P9** Describe how lifestyle choices can affect human health. Assessment activity 4.8	**2C.M7** Discuss the advantages and disadvantages of vaccination programmes. Assessment activity 4.9	

How you will be assessed

The unit will be assessed by a series of internally assessed tasks. You will be expected to show an understanding of biology relevant to the relationships between organisms and their environment, the effects of human activity on the environment and the factors that affect human health.

The tasks will be based on scenarios which place you as the learner in the position of working in a number of environmental and health-related sectors; for example, as an environmental scientist in a local council, as a health journalist or as a personal fitness trainer. Your actual assessment could be in the form of:

- a table showing the results of your investigations
- a poster to illustrate important points to patients in a GP surgery
- a magazine article.

Key term

Mutation – A change to the structure of genetic material (DNA or chromosomes) that may lead to a change in a characteristic.

Discussion point

Why do you think having blue eyes, instead of brown eyes, may be useful to people living in temperate regions where the sunlight is less intense than in tropical regions?

Link

Look back at lesson 1.4 to remember what you learnt about genes and the structure of DNA.

Did you know?

Humans have a mutation of a gene, called caspase-12, that makes them less likely to get septicaemia (blood poisoning).

You can tell the difference between a cat and a dog, and between cats, dogs and humans.

Humans all belong to the same **species** but each of us is unique and different from all other humans in some way. Some characteristics that make us different are controlled by *genes* and some by the *environment*.

We each inherit a unique combination of alleles (versions of genes) from our parents, so we are different from each other. However, we are still all members of the human species.

So there are differences *between* species and *within* a species.

Genetic variation

Genetic variation is caused by **mutations** in genes. Genes are sections of DNA found on chromosomes.

Bar-headed geese flying high over mountains.

When bar-headed geese migrate, they fly high over mountain ranges. Most birds are unable to fly so high, as there is a lot less oxygen in the air for them to breath. Bar-headed geese have a mutation in the gene that codes for their haemoglobin which makes their haemoglobin different from that of other birds. This means that bar-headed geese can absorb and carry more oxygen in their blood and so they can fly very high.

Mutations

Mutations are changes in the structure of DNA.

Some gene mutations may be harmful, but some can be useful or neutral. Having blue eyes instead of brown eyes is the result of a mutation that arose in the human population about 6000 years ago.

An example of how a mutation can be useful is the mutation to the gene for coat colour that makes the coat of arctic hares white and gives them camouflage. However, the same gene could be harmful if the climate changed and there was no snow in the arctic regions where these hares live.

Environmental variation

If you grew many cuttings from one plant they would all be genetically identical. If you grew batches of them in different types of soil (mineral content) or at different temperatures, or gave them different amounts of water, they would show variation. This is because of the differences in the plants' environment.

(a)

(b)

(c)

(d)

These vine leaves all have the same genes but show environmental variation. **(a)** Normal vine leaf, **(b)** vine leaf with deficiency of potassium, **(c)** vine leaf changing colour as chlorophyll breaks down in the autumn, **(d)** vine leaf damaged by insect pests.

Activity A

For each of the following, say whether it is caused by genes or environment:

1. you move to a new area when you are a child and you begin to speak with a different regional accent
2. the colour of your eyes
3. your blood group
4. an athlete's heart wall becomes thicker when he trains
5. being colour blind.

Link

Look back to lessons 1.5 and 1.6, for more information about genes and mutations and their inheritance.

Take it further

Some variations are caused by a combination of genes and environment. For instance, we have many genes that may determine height and many that may determine intelligence. However, anyone who is underfed or has no intellectual stimulation while growing up will not reach their genetic potential.

Lesson outcome

You should understand that the characteristics of organisms vary within and between species.

4.2 Evolution

Key terms

Evolution – Gradual change over a period of time.

Population – A group of organisms of the same species, living in the same ecosystem at the same time, and able to interbreed.

Predation – The killing and eating of one kind of organism by another kind of organism.

Natural selection – Mechanism for evolution. The best-adapted organisms survive and reproduce, and the alleles for the favourable characteristics are passed on to their offspring.

Did you know?

Starfish are distant relatives of humans. Along with sea urchins, starfish belong to a group called echinoderms (spiny skin). These animals are invertebrates, but are very closely related to vertebrates.

Pearl starfish on the sea bed.

There are millions of species of organisms on planet Earth and thousands of new ones are discovered each year. Scientists have observed that some species are closely related to other species, and species have gradually changed throughout the Earth's history. This change in species over time is called **evolution**.

There are many different ecosystems in the world, and within each ecosystem there are **populations** of organisms that are well adapted to survive.

Natural selection

Charles Darwin spent many years observing and investigating different organisms. He saw that:

- offspring look similar to their parents
- all organisms produce many young, but not all the young survive
- there is always competition between members of a population for resources, such as food, shelter and mates
- there is variation between individuals in a population, with some being better adapted than others to survive in their environment.

Both these animals are adapted to survive in their environment. They have thick, white fur to protect them from the cold climate. The white fur also acts as camouflage. The polar bear has sharp claws and teeth for killing its prey, but the arctic fox can run fast to escape.

Activity A

Think of how the following organisms are adapted to survive in their environments. You may need to look in biology textbooks or use the Internet to help you.

1. Cactus plants – live in dry areas.
2. Dromedary or Arabian (one-humped) camels that live in desert areas of North Africa.

Charles Darwin developed a **theory** to explain how new species have evolved from existing species over time. His theory of natural selection has the following main points.

- Within any population of organisms there is variation.
- The best-adapted organisms will survive because they get more resources, such as food, shelter and mates, or are better able to escape **predation**. The survivors breed and pass on these favourable characteristics to their young.
- Organisms less well adapted to their environment are less likely to survive and so have fewer offspring.
- Over many generations, the proportions of individuals with the favourable characteristics will increase and the individuals without the favourable characteristics may disappear altogether from the population.
- When individuals are different enough from, and cannot interbreed with, the original population, they are a *new species*.

As discussed in lesson 4.1, gene mutations cause the genetic variation and adaptations which drive natural selection.

Assessment activity 4.1

| 2A.P1 | 2A.M1 | 2A.D1

You are an evolutionary biologist working for a consultancy firm. You need to produce information to tell some ecologists about the causes of variation within and between species.

Use the information in this book and do some further research from books and the Internet to help you with this activity.

1 Make a table showing examples of variation between humans due to genes, and variation between humans due to the environment.
2 Describe how **(a)** genes and **(b)** the environment can bring about variation.
3 Explain how **(a)** variation caused by genes and **(b)** changes in the environment can both contribute to the process of evolution.
4 Evaluate the importance of variation due to genes and changes in the environment in the process of evolution – either causing species to become extinct or enabling them to survive.

Tips

For 2A.P1 you need to go into some detail about how changes to genes can cause variation. Think of a characteristic caused by a gene and think about the effects of different alleles of that gene on the individuals with those alleles. Now think of variation caused only by the environment and say how the difference in environment (for example amount of food eaten) causes people to vary.

For 2A.M1 you need to explain how changes to the environment can favour individuals with particular alleles and how natural selection can bring about evolution. Use an example if possible. If you go to lesson 4.14, you will find some information on the evolution of antibiotic-resistant bacteria. These bacteria have a change to their genes which makes them resistant to antibiotics.

For 2A.D1, think about which has the greater impact on whether an organism survives or becomes extinct – genes or environment. Remember that for evolution to happen the useful variation has to be inherited by the offspring.

Take it further

A recent example of natural selection is the peppered moth. There are two types of peppered moth – light coloured and dark coloured. The variation is the result of mutations in their genes.

Research, using books or the Internet, to find out how and why the ratio of dark to light coloured peppered moths has changed in the UK, since from before the Industrial Revolution to the present day.

Peppered moths.

Lesson outcome

You should know that evolution is a gradual process, involving gene mutation and natural selection, which can lead to the development of new species.

4.3 Interdependence

Get started

Do you think humans depend on plants for their existence?

Look around you. Think of all the things you have at home and in your college laboratory that come from plants. Make a list.

All living organisms rely on their environment and on other organisms for their survival.

Food chains

Food chains show what organisms eat.

The arrows in food chains show the direction of energy flow from one organism to the next. Each link in the food chain is a **trophic (feeding) level**.

- **Producers** are organisms that start the food chain because they carry out photosynthesis and produce their own food.
- **Consumers** are organisms that obtain their energy by eating other organisms or parts of them.
  1. Animals that eat plants are called **herbivores**, or **primary consumers**.
  2. Animals that eat primary consumers are called **secondary consumers** or **carnivores**.
  3. **Tertiary consumers**, or top carnivores, eat secondary consumers.

Link

Look back to lesson 1.1 to see how plant cells in producers are adapted to photosynthesise and make food.

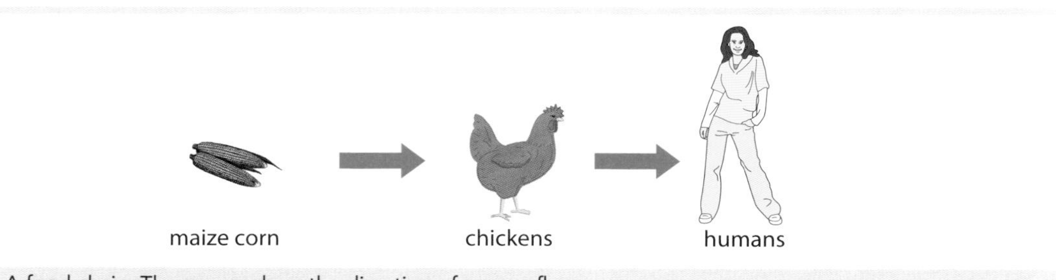

A food chain. The arrows show the direction of energy flow.

Did you know?

Most food chains start with plants or algae that can photosynthesise. However, at the bottom of the ocean there are thermal oceanic vents where there is no light. Certain types of bacteria live here and use chemical reactions to get their energy and grow. These bacteria are eaten by tube worms, clams and shrimps.

As you move from one trophic level to the next in a food chain, energy is lost during movement and as heat from **respiration**.

In any ecosystem, the number of organisms at each trophic level decreases. There are fewer secondary consumers than primary consumers and very few top carnivores.

Some organisms die without being eaten by a consumer. This dead matter is broken down by **detritivores** such as beetles and worms, and **decomposers** such as **fungi** and **bacteria**.

Activity A

In a field with dairy cows grazing grass, 30 kg of grass is eaten by cows to produce 3 kg (3 litres) of milk. Both grass and milk contain nutrients and energy. If a child drinks 3 kg of milk, 0.25 kg of the milk will be used to make new tissue in the child.

1. What percentage of the grass eaten by a cow becomes milk that can be drunk by a child?
2. What percentage of the nutrients and energy in milk will become new tissues (e.g. bone, muscle, blood, skin) in the child?
3. What percentage of the nutrients and energy in the grass become tissues in the child?
4. What has happened to the rest of the energy?

Food webs

Most animals eat, or are eaten by, many other types of organism. A food web shows this variety of food sources and the **interdependence** of organisms within an ecosystem.

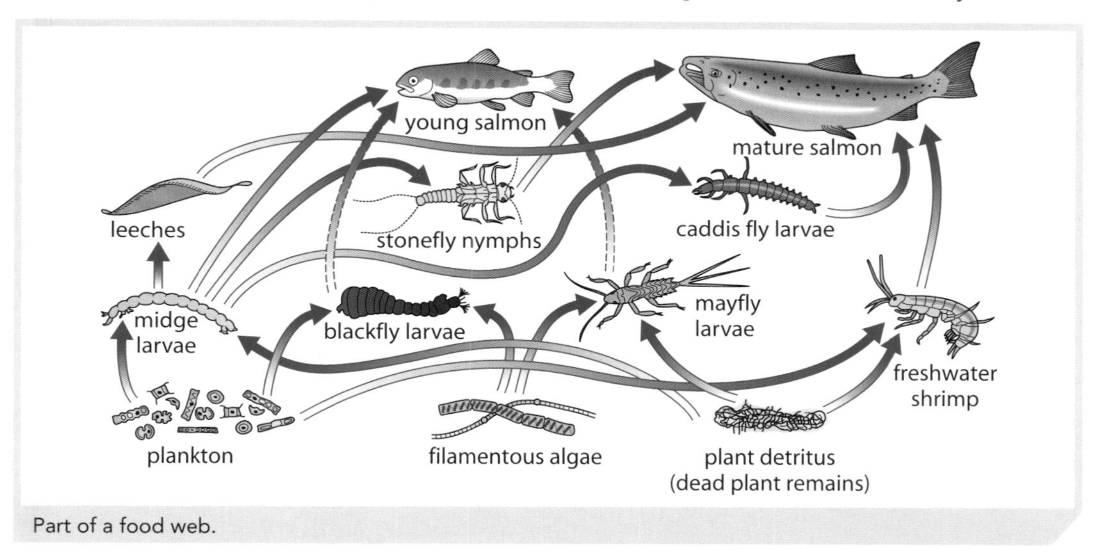

Part of a food web.

Predator–prey relationships

A **predator** is an animal that hunts other animals (**prey**) for food.
The relationship between predator and prey populations is often shown in a graph.

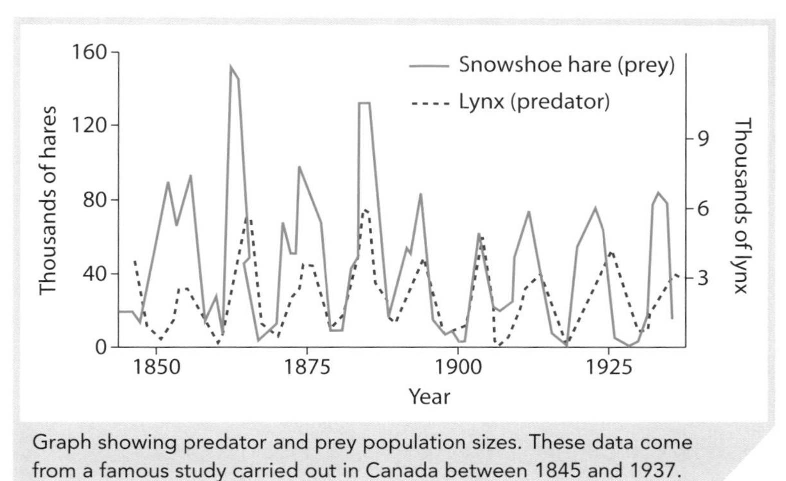

Graph showing predator and prey population sizes. These data come from a famous study carried out in Canada between 1845 and 1937.

Lynx with its prey – a snowshoe hare.

- As the predator population increases, more prey are eaten.
- The prey population then decreases so there is less food for the predators.
- The predator numbers fall as they have less food so fewer predators survive to reproduce.
- As there are fewer predators, fewer prey animals are eaten, so more prey animals survive and reproduce.
- There are more prey animals for predators to eat, so predator numbers increase and the cycle starts all over again.
- Prey populations may also drop if their food source decreases, for example due to cold weather. So prey numbers can determine predator numbers.

Case study

Luke is an environmental scientist. He is involved in a study about the reduction of coral on the Great Barrier Reef. Other scientists have already proposed a hypothesis that the coral is being overeaten by a type of starfish. Luke went to the Great Barrier Reef and collected data about the food chains. He found that coral eat plankton, starfish eat coral and the top carnivore, a big fish called wrasse, eat the starfish. He also found that humans were overhunting the wrasse. This meant that more starfish were surviving and overeating the coral. His findings supported the evidence gathered by other teams of scientists.

1 What advice do you think Luke should give to the Australian Government's Department for the Environment so that they could develop a policy to prevent this loss of coral from the Great Barrier Reef?

2 Why do you think it is important that other teams of scientists gather the same evidence?

Assessment activity 4.2

| 2A.P3 | 2A.M2

You are a visiting speaker at a local school and have been asked to explain food chains and food webs.

1 Look at the diagram of a food web on page 165. Write down three food chains that you can see in this food web. Your food chains should each have four organisms in them. In each case, label the producer, primary consumer, secondary consumer and tertiary consumer.

2 Make a large poster to show the different ways in which some of the organisms in the food web show interdependence (depend on each other).

3 Discuss the factors that affect the relationships between the organisms you have referred to in part 2.

Tips

For 2A.P3 think of as many ways as possible that the organisms rely on each other. Show how characteristics of producers, herbivores (primary consumers) and predators (secondary consumers) determine their place in the food chain. For 2A.M2 think about all the factors that could increase or decrease the numbers of the organisms.

Lesson outcome

You should know that interdependence of organisms can be shown using food chains, food webs and predator–prey relationships.

WorkSpace

Jason Smith

Environmental Consultant

I work for an independent, not-for-profit company, specialising in advising businesses and the public about reducing their carbon emissions. For example, we may advise businesses to:

- reduce the temperature in their buildings, and so use less fuel
- make sure employees turn off computers and lights at night
- make their buildings more energy efficient
- allow employees to work from home sometimes to reduce their travel to and from work
- hold meetings with colleagues abroad by video-link rather than flying out to meet them.

We may advise individuals to:

- share cars, use public transport or walk/cycle for short journeys
- shop once a week or shop online
- plan journeys so they don't get lost and can avoid congested areas
- switch off engines if they are stationary for more than 2 minutes
- drive steadily, avoiding harsh accelerating or braking
- not carry heavy objects in the boot of the car
- have cars regularly serviced and choose a fuel-efficient type of vehicle.

We can calculate an individual's or a company's carbon footprint and encourage them to buy carbon credits to offset their carbon footprint. This means they pay for producing emissions, and the money is used to:

- plant more trees along roadsides or in school grounds – the trees absorb carbon dioxide from the air
- help develop and subsidise use of solar panels or solar cookers in some developing countries
- help develop wind energy plants
- help introduce energy efficiency into hotels in India.

Think about it

1. Why do you think it is important to try to get people to reduce their carbon emissions?
2. How do you think you can convince people that their individual small efforts are important?

4.4 Classification

Get started

How many different species of living things can you name in 5 minutes? Make a list. Did you remember to include living things other than plants and animals?

It is difficult to say exactly how many species there are on Earth, but it is probably around 15 million, with new ones being discovered every week. New species are classified by biologists. This involves looking at the new organism and comparing its characteristics to the characteristics of known species.

The five kingdoms

Living organisms are grouped according to their similarities and differences:

- similar *species* are grouped into the same *genus*
- similar *genera* into the same *family*
- similar *families* into the same *order*
- similar *orders* into the same *class*
- similar *classes* into the same *phylum*
- similar *phyla* into the same *kingdom*.

The characteristics of the five kingdoms are shown in the table.

Bacteria (prokaryotes)	Protoctists	Fungi	Plants	Animals
	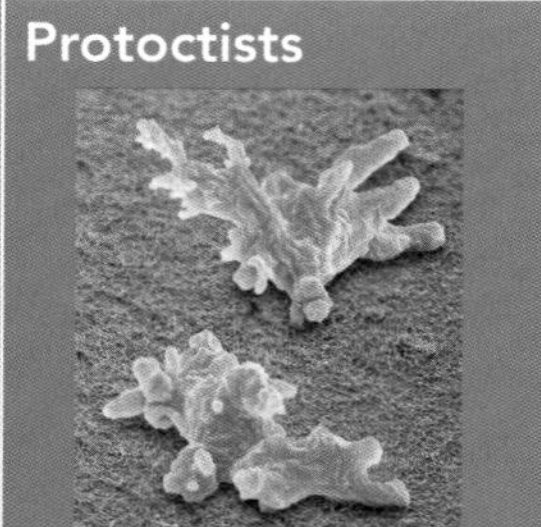			
Very small, made of one cell Cells have non-cellulose walls No nucleus – DNA is free in cytoplasm Many are useful as agents of decay A few can infect us and make toxins	Eukaryotic organisms Many are unicellular Some can photosynthesise Larger than bacteria but still microscopic	Eukaryotic Multicellular Non-cellulose cell walls Some live in or on their food (parasitic) and can cause diseases of crop plants, livestock and humans Some feed on dead matter	Eukaryotic Multicellular Cells have walls made of cellulose Photosynthesise Include mosses, ferns, trees and flowering plants	Eukaryotic Multicellular Cells do not have walls

Within the animal kingdom, animals can be grouped into those with backbones (vertebrates) and those without a backbone (invertebrates).

Vertebrates

Vertebrates have two main features:

- a spinal cord and brain
- a vertebral column (backbone).

The table shows the main characteristics of the five vertebrate classes.

Fish	Amphibians	Reptiles	Birds	Mammals
Live in water Scaly skin	Live on land or water Moist skin	Dry scaly skin	Have feathers	Have hairy bodies
Use gills to obtain dissolved oxygen from water	Use gills when young to obtain oxygen from water, and later use lungs to obtain oxygen from air	Have lungs for obtaining oxygen from air	Obtain oxygen from air using lungs	Obtain oxygen from air using lungs
Body temperature may vary with external temperature	Body temperature may vary with external temperature	Body temperature may vary with external temperature	Maintain constant body temperature	Maintain constant body temperature
Most use external fertilisation and lay eggs	Return to water to breed Most use external fertilisation and lay eggs	Internal fertilisation Lay eggs with leathery shells	Internal fertilisation Lay eggs with hard shells	Internal fertilisation Give birth and feed young on milk made in mammary glands

Where do viruses fit in?

Viruses do not fit into the above system as they are not really living. Viruses are not made of cells. They cannot reproduce on their own – they have to invade a living cell and hijack that cell's structures to make copies of the virus.

Newt tadpole – note the gills on this young amphibian.

Activity A

1. How are reptiles and birds similar? How are they different?
2. When bacteria and fungi were first discovered they were classified as plants, because at that time scientists thought all living things were either plants or animals. Now bacteria have their own kingdom. Use your knowledge of cell structure (as outlined in lesson 1.1) to discuss why you think **(a)** bacteria and **(b)** fungi are no longer classified as plants.

Just checking

Complete this table for the classification of humans, cats and dogs.

	Cat	Dog	Human
Kingdom	Animal		
Phylum	Chordate	Chordate	Chordate
Sub phylum		Vertebrate	
Class			Mammal
	Carnivore	Carnivore	Primate
	Felidae	Canidae	Hominidae
	Felis	*Canis*	*Homo*
Species	*catus*	*familiaris*	*sapiens*

Lesson outcome

You should know that organisms are classified according to their characteristics.

4.5 Keys

Get started

How many different species of living organisms do you think there are on Earth? Where do you think they have all come from?

A key is a way of identifying an unknown organism. Scientists can use keys to classify and sort living organisms that have similar characteristics into groups, so they can easily identify them.

Grass snakes are a protected species.

Case study

Jane is an ecology consultant. She is often called in by councils when they are carrying out a development project such as building a road. Her job is to identify the plant and animal species in the area and see if any of them are an endangered or protected species.

It is illegal to harm a protected species or disturb their habitat, so the species would have to be moved or their habitat protected. For example, if there was a great crested newt or a grass snake in an area to be developed, losing that habitat would have a major impact on that species. Sometimes Jane recommends to councils that they must close a road – such as when toads are crossing to reach their breeding grounds.

1 Explain how building a road in an area where grass snakes live may affect the population of grass snakes.

2 Squirrels, cats, hawks, owls and weasels eat grass snakes. Grass snakes can swim, and live in areas where there is water as they eat toads and frogs. They also eat mice, worms, small fish and insects.

   Use this information and draw a food chain and food web for a habitat where grass snakes live. Think about what might happen to the numbers of other animals in the habitat if grass snakes were killed.

A branching key

A branching key asks a series of questions. Each question has two answers, often 'yes' or 'no', and the answer to each question leads you to another question. Eventually, the answers will lead you to the name of the specimen.

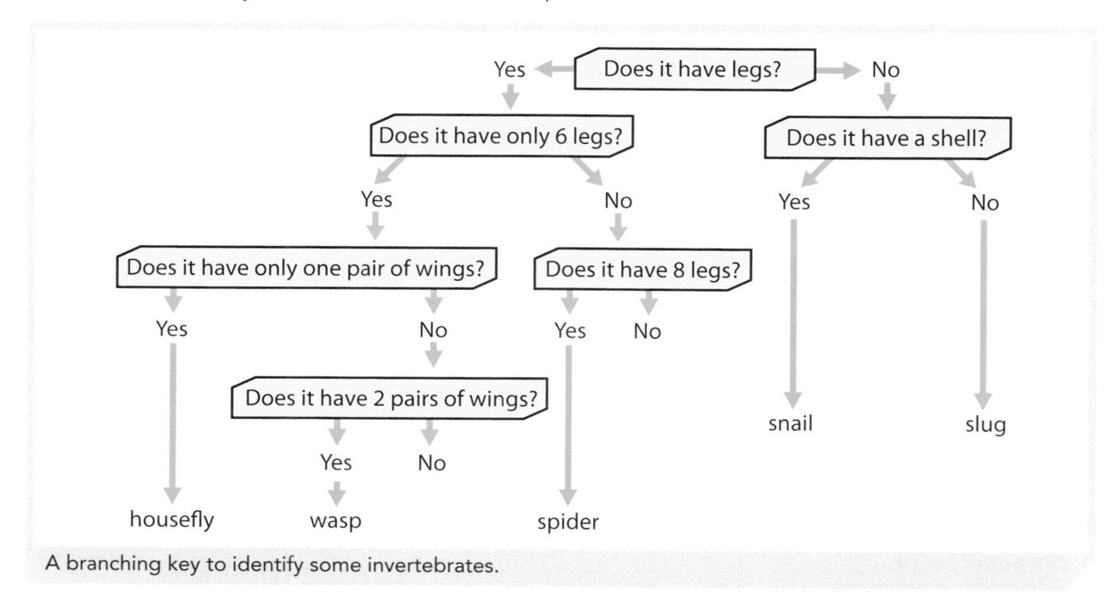

A branching key to identify some invertebrates.

A numbered identification key

The numbered key below also allows you to identify organisms.

1	Has wings Does not have wings	Go to 2 Go to 3
2	Has 1 pair of wings Has 2 pairs of wings	Housefly Wasp
3	Has a shell Does not have a shell	Snail Go to 4
4	Has legs Does not have legs	Spider Slug

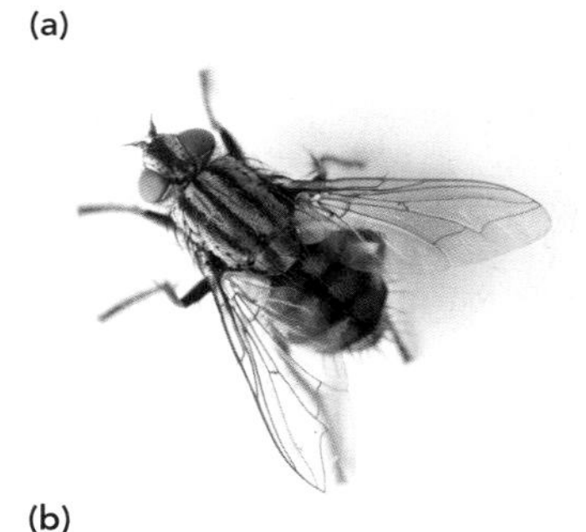

(a) Housefly, *Musca domestica*, has one pair of wings; **(b)** wasp has two pairs of wings.

Activity A

(a) (b) (c) (d) (e)

1 Use the branching key to identify animals **a–e**.

2 Use the numbered key to identify animals **a–e**.

3 Which key did you find easier to use?

Assessment activity 4.3

| 2A.P2

You are an experimental biologist helping a wildlife park to make an identification key to be used by primary school children when following a nature trail.

1 Select six organisms and create identification keys so that someone else can identify them. Draw the organisms or paste in pictures of them.

(a) Use the branching yes/no method.

(b) Use the numbered method.

2 Make a poster or chart to show how scientists classify organisms.

Tips

To achieve 2A.P2 you need to describe the main groups of living things and explain what the members of those groups have in common. These are the characteristics that are used to classify these organisms into a particular group.

You can illustrate your chart with pictures or drawings of organisms.

Lesson outcome

You should be able to construct and use keys to show how organisms can be identified.

4.6 Agriculture and ecosystems

Key terms

Ecosystem – All the organisms living in a particular area and the area itself. The organisms interact with each other and with the non-living features of the area.

Habitat – The place where an organism lives.

The human population

Humans originated on the African continent. About 1 million years ago they began to migrate to other parts of the world.

Farming began about 10 000 years ago in an area now called the Middle East. The practice then spread to many other areas. About 400 years ago the human population began to increase *steadily* as farming methods improved. Then, about 150 years ago the human population increased *sharply* due to:

- improved hygiene and sanitation
- improved food production and diet
- better health care
- lower infant mortality (death rate)
- more people surviving to adulthood and reproducing.

Did you know?

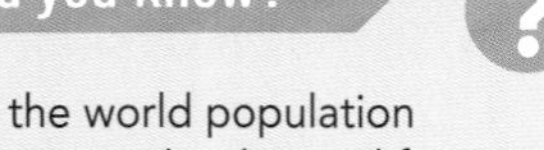

As the world population increases, the demand for meat increases. Cows have to be fed grain, so as the world population increases by 50%, the demand for grain could double.

Scientists have grown cow stem cells into muscle tissue, in a laboratory, and made protein. In the future, this may be the source of beef burgers.

How does farming affect ecosystems?

As human populations grow, more food is needed to feed everyone. Modern farming is efficient but affects the environment, as it:

- takes up space, uses resources and produces pollution
- uses chemical fertilisers made in factories that burn fossil fuel; burning fossil fuel contributes to climate change
- uses machinery, e.g. tractors, that burns fossil fuel
- uses pesticides (insecticides and herbicides) to reduce loss of food due to it being eaten by pests; these pesticides may accumulate in the environment and in food chains and harm other organisms
- produces large amounts of animal waste (slurry) that can cause pollution
- rears many cows that produce methane which contributes to climate change
- reduces genetic diversity when just one type of crop is grown (monoculture).

Large area of just one type of crop – rice – being harvested in the USA. Growing rice requires large amounts of water.

Deforestation

Cleared forest in South America.

Large areas of forest are cleared to grow crops to feed people or livestock, or to grow grass for livestock to graze. Forests are also cleared for timber for fuel, building materials or making furniture. This is called **deforestation**.

- Deforestation removes trees which then means less carbon dioxide is removed from the atmosphere.
- More carbon dioxide in the atmosphere contributes to climate change.
- Clearing forests means that **habitats** for wildlife are lost.
- Valuable resources are lost as trees provide wood and many are a source of medicines.
- Lack of tree roots to hold the soil together leads to soil erosion. Landslides and mudslides may then occur, burying villages and killing people and wildlife.
- Water and minerals may drain quickly from the soil, so fewer minerals are available for plant growth.

Discussion point

In the future we may have vertical farms. This will save space. They could be built within cities and would consist of multi-storey greenhouses.

This would allow many crop plants to be grown but not to cover such a wide area of the Earth's surface.

The conditions inside would be controlled so fewer pesticides would be needed, as there should be fewer insect pests. The cities' sewage could be used as fertiliser.

In small groups, discuss the advantages and disadvantages of vertical farms.

Activity A

Farming doesn't only happen on land. The amount of farmed fish and shellfish in the world has more than doubled in the last 15 years. Some people think this is better than fishing wild fish from the seas, but this is not really true.

A fish farm in Greece.

- The farmed fish have to be fed on fishmeal which is made from wild fish. For every 1 kg of farmed fish produced, between 1.9 kg and 5 kg of wild fish are used to make their food.
- Crowding of fish raised in farms causes diseases to spread amongst them.
- Antibiotics used to treat them end up in the water and get washed into the surrounding sea or rivers.
- Other waste products, such as fish faeces, pollute water surrounding the fish farm.
- If farmed fish escape they may disturb the gene pool of their wild relatives by interbreeding with them.

1 In small groups try to think of some arguments FOR fish farming.

2 Why do you think 1.9–5 kg of wild fish have to be made into feed to raise just 1 kg of farmed fish? (Hint: think back to food chains.)

3 Why do you think it is more efficient for humans to eat fish caught in the wild and not farmed fish?

Lesson outcome

You should understand how human activities alter ecosystems through deforestation and agriculture.

4.7 Transportation and ecosystems

Get started

Make a list of all the food you ate yesterday. Now find out where all the ingredients come from. For example, bread is made from flour, mainly from wheat, which is grown in the UK. Burgers are made from minced beef, which probably comes from cows reared abroad, possibly as far away as South America.

Food miles

A food mile is the distance food travels from where it is grown or produced, to where it reaches the consumer. This may include:

- moving livestock from farm to slaughterhouse, and from a processing plant to a shop
- moving fish between countries to be processed and then distributing them to supermarkets
- moving farm produce to local markets
- moving fruits and vegetables to other countries
- moving ingredients to processing plants, some of which may be in other countries
- trips made by consumers in their cars to shops.

Calculating food miles helps us understand the actual environmental impact of producing our food.

- Transport and refrigeration need fossil fuel and release carbon emissions as well as other pollutants.
- Carbon dioxide and oxides of nitrogen are greenhouse gases and contribute to climate change.
- Sulfur dioxide and nitrogen oxide, produced when fossil fuels are burned in vehicle engines and in power stations, contribute to acid rain.

Link

Look at lessons 1.15, 1.16 and 4.10 for more information about acid rain.

Food miles only contribute about 12% of the total carbon emissions produced by farming – the rest are produced during the actual farming process.

Activity A

These are the percentage (%) carbon emissions produced by the different ways of transporting food.

- By road: 60%
- By air: 20%
- By rail: 10%
- By sea: 10%

Draw a suitable graph to illustrate these figures.

Discussion point

How do you think farming contributes to carbon emissions?

It may seem environmentally harmful to carry truckloads of tomatoes from Spain to the UK, but it adds *less* of a carbon footprint than growing tomatoes in heated greenhouses in the UK.

The fruits and vegetables in this shop were grown in many different countries.

These trucks burn fossil fuel and emit carbon dioxide while transporting fruit and vegetables long distances.

Did you know?

Each year the UK produces 19 million tons of carbon dioxide during the transportation of food.

Human travel

More and more people take holidays abroad and many people also travel for their work. Air travel cuts travelling time but produces more carbon emissions per person than other forms of travel.

Activity B

If people travel from London to Paris by rail, the amount of carbon dioxide generated on the journey is 22 kg per person on the train. If the same number of people went from London to Paris by air, the amount of carbon dioxide generated on the journey would be 244 kg per person on the plane.

Calculate the percentage increase in carbon emissions by flying from London to Paris, rather than going by train.

Building roads, railways or airports uses up land that cannot then be used to grow food. Many of the building materials, plus the train carriages and engines, cars and planes, need to be manufactured using fossil fuel.

Just checking

1 Jane and Usha had roast lamb for lunch. Usha's lamb came from Wales and Jane's from New Zealand. Comment on their two meals in terms of the food miles and carbon emissions.

Lesson outcome

You should understand that transportation, of food and for travel, affects ecosystems.

4.8 How fertilisers affect ecosystems

Key term

Eutrophication – Poisoning of areas of water (streams, rivers or lakes) by excess nitrates and phosphates, which cause algal bloom, death of plants, depletion of oxygen and death of fish.

For healthy growth, plants need minerals from the soil, as well as light and water for photosynthesis. The table lists some of the minerals needed by plants, and why.

Mineral	Why plants need it
Magnesium, Mg	To make chlorophyll needed for photosynthesis
Nitrogen, N, in the form of nitrates or ammonium compounds	To make amino acids which are needed to make proteins for growth To make new cells and organelles, new leaves and enzymes
Phosphorus, P	To make cell membranes and substances within cells
Potassium, K	To open leaf pores to let in carbon dioxide for photosynthesis To activate many plant enzymes involved with photosynthesis, respiration and growth

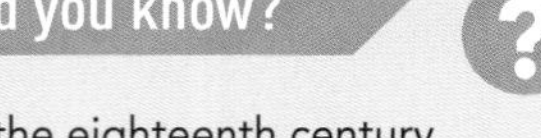

Did you know?

In the eighteenth century, guano (faeces of seabirds, bats and seals) was the most widely used fertiliser.

Link

The Haber process for the production of nitrogen-based fertiliser is discussed in lesson 2.11.

Organic and inorganic fertilisers

Farms used to be small and mixed – raising both crops and livestock. The manure from animals was added to soil to improve crop growth. Unusable parts of the plants were also composted and added to soil as fertiliser. Human sewage, crushed animal bones, blood from slaughterhouses and bonfire ash can also be used as a fertiliser. These are all *organic* fertilisers.

Since the human population has increased rapidly, farming has to be more efficient. Farmers need to grow more crops on less land, so organic fertilisers are not enough to give the necessary increases in yield.

By the 1940s, scientists had discovered how to make *inorganic* fertilisers very cheaply. This led to an increase in the use of fertilisers containing the minerals needed by plants and an increase in crop yield. However, overuse of fertiliser causes environmental problems.

Eutrophication

Eutrophication occurs when too much fertiliser, whether organic or inorganic, is used.

- When it rains, excess fertiliser is washed off the land into streams, rivers or lakes – this is known as leaching.
- The extra nitrates and phosphates encourage algae in the water to grow more than usual (algal bloom).
- The algae cover the surface of the water and prevent light reaching plants below the surface.
- These plants die and bacteria in the water decompose them.
- These bacteria use oxygen from the water for their aerobic respiration.
- Fish cannot get enough oxygen for their respiration and they die.

Dead fish in a river, due to lack of oxygen caused by eutrophication.

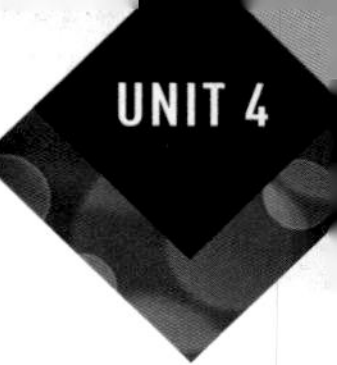

Activity A

During the Second World War, old pastureland and lawns were often dug up so that more vegetables could be grown. Research, using text books and the Internet, how this could have contributed to eutrophication.

Posters like this one were seen in Britain during World War II, encouraging people to grow their own food.

Case study

Aaron is a laboratory technician who works for one of the UK's providers of water testing services.

Samples come into the laboratory from field study officers who take them from rivers, streams and lakes. One of the tests Aaron does is a BOD – biological oxygen demand test. Oxygen levels in samples are measured at once, and then after being kept for 5 days in the dark. If the levels of oxygen have decreased this indicates that organisms in the water have made a high demand on that oxygen. A high BOD indicates that there are high numbers of aerobic (oxygen-needing) bacteria in the water sample. This indicates a high level of pollution by nitrates – either from fertiliser or sewage.

Environmental biologists can then work with farmers and advise them on how to reduce eutrophication. For example, the farmers can:

- test their soil for minerals to see how much fertiliser they need to add
- only add fertiliser when a crop is growing so that the plants absorb the minerals
- avoid adding fertiliser when it is raining
- leave a strip of land around the edge of a field, next to a **watercourse**, untreated
- check that their slurry pits are not leaking and, if they are, repair their liners.

Aaron also tests waste waters from industrial plants to make sure they do not contain pollutants. Companies have to comply with strict regulations to prevent such pollution, which could damage the environment.

1 Why do you think the samples being tested have to be kept in the dark for 5 days, before being re-tested?

2 How do you think authorities can make sure that companies do not add pollutants to water?

Just checking

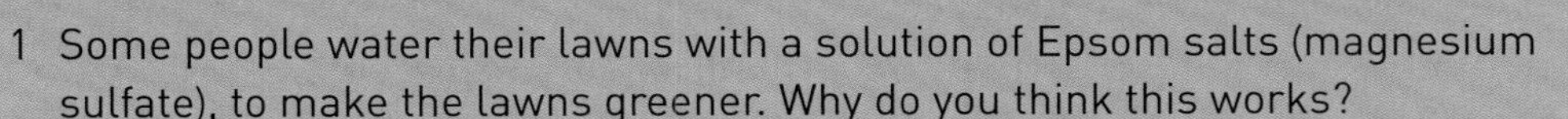

1 Some people water their lawns with a solution of Epsom salts (magnesium sulfate), to make the lawns greener. Why do you think this works?

2 Why do you think (**a**) more nitrates and (**b**) more phosphates in water encourage growth of algae?

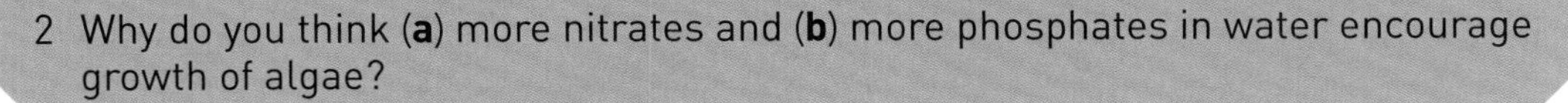

Lesson outcome

You should understand that overuse of fertilisers causes eutrophication.

4.9 Pesticides and ecosystems

Key terms

Bioaccumulation – The accumulation of a substance in the tissues of living organisms.

Pesticide – Substance that can kill an organism considered to be a pest.

Herbicide – A type of pesticide that kills weeds.

A lot of the food that farmers grow would be eaten or destroyed by pests, such as insects, birds and snails, if we did not use **pesticides**. Weeds are also pests as they compete with crop plants for space, light and water.

Use of DDT

During the 1940s, DDT was developed to control malaria and typhus fever among soldiers in the tropics. After the Second World War, DDT was used in agriculture to kill insect pests.

The harmful side effects

By the early 1960s, scientists found that many predatory birds were not reproducing properly due to the high amounts of DDT accumulating in their tissues and body fat.

- DDT kills useful insects as well as harmful ones.
- DDT stays in the insects' tissue and when predators eat these insects the DDT accumulates in their tissues.
- DDT also persists in soil and water and can be absorbed by fish.
- Animals at the top of a food chain, such as eagles, build up a lot of DDT in their tissues. This is called **bioaccumulation**.
- DDT enters the human food chain as it persists in soil and water and may cause cancer in humans.

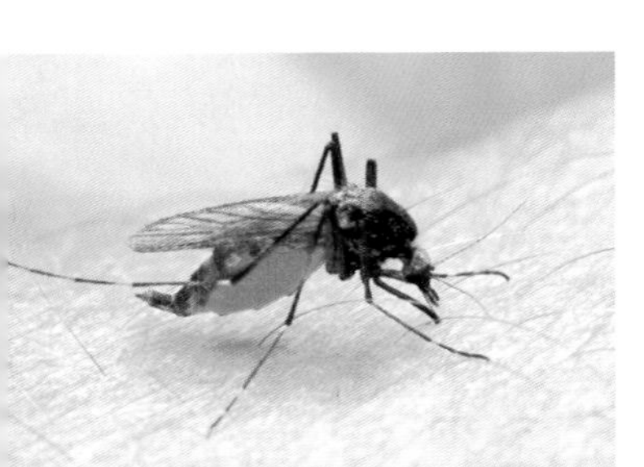

Female mosquito taking a blood meal and spreading malaria.

By the 1970s, the use of DDT in agriculture was banned worldwide. But it is still used in some parts of the world to kill mosquitoes and reduce the spread of malaria.

Pesticide spraying. Each year, agricultural workers are exposed to many pesticides, with about 18 000 workers dying worldwide.

Take it further

Honey bee numbers have been declining because of something called colony collapse disorder. Scientists think this may be linked to the use of a new type of pesticide called neonicotinoid. It does not directly kill bees but it affects their nervous systems and stops them keeping themselves and their hives clean. This means they are more likely to be infected by parasites. The use of this pesticide has been restricted in some countries.

Interfering with food chains

Other pesticides are used by farmers and, although they may not bioaccumulate, they can disrupt food chains. If some insects are killed, animals that feed on those insect species may starve if they cannot find other food sources. Some insecticides may also kill useful insects such as bees, which are very valuable because they pollinate many species of plants and produce food (honey) which humans can eat.

Activity A

1. Populations of insects can become resistant to a pesticide. Explain how this happens. (Hint: see lesson 4.2 for information on how natural selection works.)
2. Why do you think farmers apply different pesticides alternately, rather than just keep on applying the same type of pesticide?

Herbicides and the environment

Herbicides are used to kill weeds. Most herbicides interfere with photosynthesis or with the growth of the plant.

- *Selective* herbicides only kill certain types of plants – so can be used in fields of crops.
- *Non-selective* herbicides kill all plants and are used to clear waste ground or railway lines.

If we remove weeds we may also remove food or habitats for wildlife, such as birds or useful insects. This means that herbicides can disrupt food chains and reduce biodiversity – the variety of living organisms within an ecosystem.

Did you know?

Some plants, such as the walnut tree, produce their own herbicides to reduce competition from other plants.

A farm overtaken by a species of weed in Australia.

Assessment activity 4.4

| 2B.P4 | 2B.M3 | 2B.D2

You are an environmental science technician investigating the effects of human activities on ecosystems.

1. Choose an ecosystem such as a wood, a lake, open countryside or the sea. Make a table showing the human activities that can affect this ecosystem.
2. Choose two more types of ecosystem. Think of all the ways that human activities affect ecosystems. Describe how these different human activities affect the ecosystems you have chosen.
3. Analyse the effects of pollutants on the ecosystems you have chosen.
4. Explain the long-term effects of pollutants on living organisms and on ecosystems.

Tips

For 2B.P4 you might like to make a separate table for each ecosystem you choose and show *how* the polluting effects of the activities harm living organisms and the ecosystems.

For 2B.M3 use data you obtain from your own practical work as well as other data you find during research. You could consider, for example, advantages and disadvantages of using DDT to combat the spread of malaria. Consider also how not using pesticides would affect our food production.

For 2B.D2 think about eutrophication. Think also about long-lasting and 'knock-on' effects, such as removing bees from the ecosystem. How would this affect our food production?

Lesson outcome

You should know that toxic herbicides and pesticides can bioaccumulate and disrupt food chains on land and in water.

Living indicators

Some animal and plant species can indicate how polluted an area is.

Lichens

A lichen is made up of two organisms: a fungus and an alga. The fungus provides stability and structure, and traps water, while the alga contains chlorophyll and so provides food by photosynthesis. Different types of lichen grow on trees, fences or stone surfaces.

Lichens are slow-growing and live for a long time. They are sensitive to sulfur dioxide in the air, but some types are more sensitive than others. The higher the levels of sulfur dioxide, the fewer the types of lichen species there will be in that area. However, they are not present in all the areas we may want to test.

Crusty lichen on stone. This type can grow in air that has a high level of sulfur dioxide.

Leafy lichens on a tree trunk.

Shrubby lichen on a tree. This type only grows in unpolluted air.

Freshwater shrimps

Bacteria use oxygen when they decompose organic pollution (sewage and fertilisers) in rivers and streams.

Some freshwater organisms, such as freshwater shrimps, can only live in well-oxygenated water, while others survive in water with low oxygen levels. So, the species present in rivers and streams can give us an indication of how clean or how polluted the water is.

Diatoms – single-celled planktonic algae.

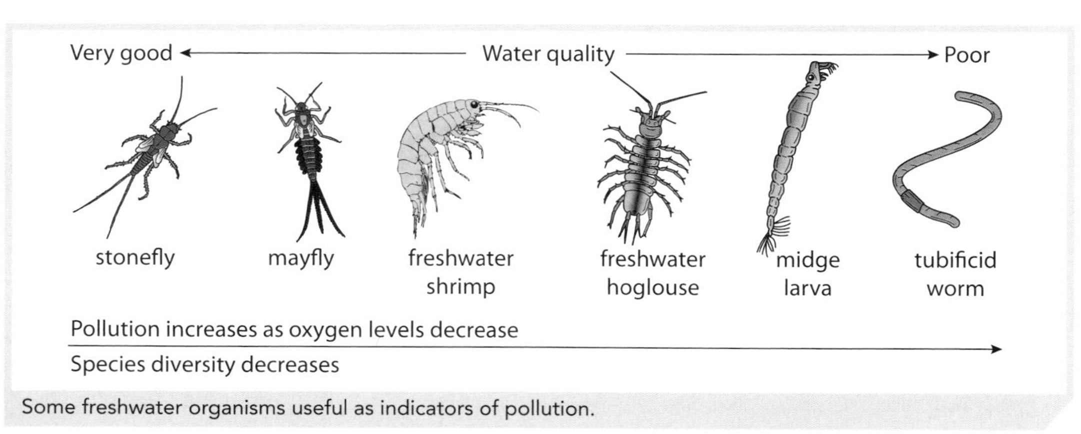

Some freshwater organisms useful as indicators of pollution.

Algae

Algae also act as indicators of pollution.

- **Algal blooms** in watercourses indicate high levels of organic or inorganic pollution.
- More filamentous (thread-like) algae coupled with fewer planktonic (single-celled) algae indicate that the water is more acidic than usual.

Clumps of filamentous algae growing in a polluted ditch.

Non-living indicators

You have already learnt about eutrophication. If water contains a lot of dissolved nitrates then algae flourish, blocking light from plants. The plants die and bacteria use oxygen to decompose them. So high nitrate levels and low oxygen levels in water are non-living indicators of pollution by sewage or fertilisers.

High nitrate levels in drinking water may harm babies.

Erosion of limestone buildings

Burning fossil fuels in power stations or in vehicles may cause acid rain. This can kill fish and trees, and also erodes limestone buildings. Acid rain occurs because:

- there is sulfur in the fuel and it joins with oxygen in air to make sulfur dioxide
- the carbon in fuel combines with oxygen to make carbon dioxide
- these gases dissolve in rainwater to make acid rain.

Acid dissolves calcium carbonate (limestone). Limestone buildings are gradually eroded by acid rain. This is seen most easily in carvings on buildings or on statues. The greater the erosion, the more sulfur dioxide is in the air.

Case study

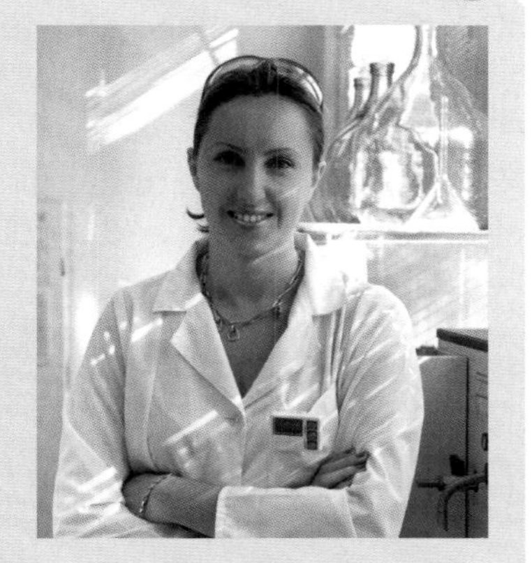

Alice is an ecologist, working for an environmental agency. She monitors streams and rivers to see which organisms are present. This tells her whether the water is more polluted than it should be. She also measures the pH of water samples. If the water is polluted, her report will be used to help the agency find the source of pollution. The agency then asks the company causing the pollution to reduce its discharges into rivers. If the company does not do so, a magistrates court can impose large fines or a prison sentence on the directors of the company.

The environmental agency works to keep rivers and wetlands clean for wildlife and for recreational use, such as boating and fishing.

1 Alice's data may have to be used in court so they have to be near the true value and must not have any errors made in the sampling and taking of readings. How do you think she can try to make sure she collects such data?

Assessment activity 4.5 | 2B.P5

You are writing a handbook for environmental scientists to use.

1 Make an illustrated wall chart that shows living and non-living pollution indicators.

2 For each indicator, show the type of pollution they measure and describe how they are used to measure the levels of the pollutants.

Link

Look back to lesson 4.8 to read about eutrophication.

Erosion of a limestone pillar of a building by acid rain.

Activity A

Use books and the Internet to find out how too much nitrate in drinking water can harm babies.

Tip

Remember that living indicators are animal and plant species. Non-living indicators are not organisms.

For 2B.P5 indicate how the pollutants are measured, using the indicators you have named.

Lesson outcome

You should understand how living and non-living indicators can be used as a measure of the level of pollution in an ecosystem.

4.11 Reducing the effects of pollution

Get started

Most local councils encourage householders to recycle as much of their rubbish as possible by providing different recycling bins and boxes for different items. List all of the items of rubbish in your home that can be recycled.

The three 'R's

We can all do something to help reduce the impact of pollution on ecosystems by remembering the three 'R's:

- reduce
- reuse
- recycle.

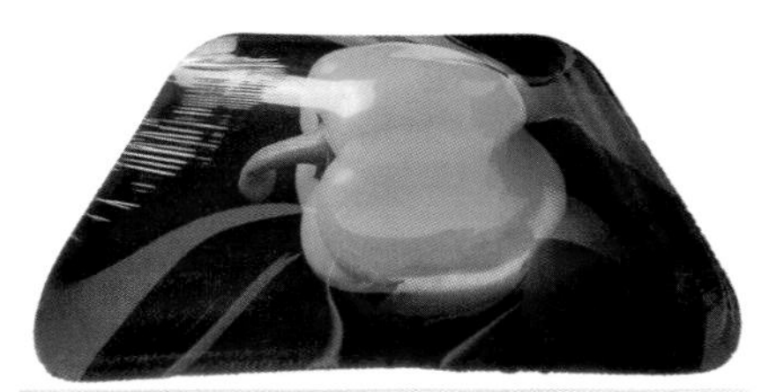

This pepper does not need to be shrink-wrapped.

Reduce

When shopping, try to buy things that aren't wrapped in lots of packaging. Not only is it a waste of plastic, it also took energy and water to make the wrapping products and the packaging is not always biodegradable.

Reuse

As well as reusing plastic carrier bags,

- toys, books, clothes and spectacles can be taken to charity shops or sent to hospitals or less economically developed countries (LEDCs)
- computers and mobile phones can be refurbished and sent to LEDCs, and can also be dismantled to reuse some of the metals that are inside them
- ink cartridges can be refilled.

This saves resources and prevents landfill sites being filled with highly toxic metals, such as cadmium, lead and beryllium from mobile phones.

Activity A

Find out how iron and steel are recycled.

Recycle

Recycling saves natural resources and reduces the amount of waste produced. It may also use less energy and produce less carbon dioxide than making new items from natural resources.

Item	Can be recycled into
Wood, e.g. old furniture	Fuel or wood shavings used for horse and pet bedding; chipboard
Tyres	Shoe soles; carpet underlay; sports and playground surfaces; boat fenders; crash barriers
Aluminium, e.g. drinks cans, foil	New cans and foil – this saves use of the raw material (bauxite) and produces less carbon dioxide than extracting aluminium from ore
Plastic	Buckets; traffic cones; new drinks containers
Glass	Wall tiles; bowls and glasses made from recycled glass
Cooking oil	Bio-diesel
Clothes and rags	Wool can be recovered to make other clothes such as uniforms
Kitchen and garden waste	Bacteria break down the organic matter and turn it into compost which we use as garden fertiliser
Metals	Jewellery; steel is recycled for many uses including in construction and car industries; iron is recycled and used to reinforce concrete

Did you know?

- 25 used 2-litre drinks bottles can be recycled into an adult-sized fleece.
- Recycling one glass bottle saves enough energy to power a TV for 15 minutes.
- Each UK household produces 1 tonne of rubbish per year; we each throw away our own body mass of rubbish every 7 weeks; every 8 months the UK produces enough waste to fill Lake Windermere.
- The amount of Christmas wrapping paper discarded each year in the UK could cover an area the size of Guernsey.

Activity B

Recycling paper produces less carbon dioxide than making paper from wood, but it uses a lot of water and produces poisonous chemicals, called dioxins, when the ink is removed. New paper is made from softwood trees that are replanted regularly, and young growing trees absorb a lot of carbon dioxide from the atmosphere. Old paper can be shredded to make pet bedding or compost.

1 In small groups discuss the relative merits of recycled versus non-recycled paper.
2 Write down three ideas for and three ideas against recycling paper.
3 Share your ideas with the rest of your class.

Discussion point

Why is it helpful to the environment if human activities produce less carbon dioxide?

Conservation

Conservation biology is about protecting and managing habitats and wildlife so that the environment is not spoilt for future generations and can still generate resources.

There are laws, government departments such as DEFRA (Department for Environment, Food and Rural Affairs) and the Forestry Commission, and other organisations such as the RSPB (Royal Society for the Protection of Birds) that help protect the environment and local ecosystems.

The Forestry Commission

The Forestry Commission manages about 1 million hectares of forest land in the UK. It is responsible for replacement planting of trees to maintain and expand forests and woodland, and for ensuring responsible harvesting of timber. This helps protect habitats for wildlife and creates areas of recreation for people.

Breeding programmes

Climate change and habitat loss have brought some species close to extinction. Such rare or endangered species can be bred in captivity (zoos or wildlife parks) and maintained there, or some of the offspring can be reintroduced into the wild. There are sometimes problems with this approach, such as inbreeding and a reduced gene pool.

This herd of rare deer at Longleat is part of a captive breeding programme.

Discussion point

Why do you think it is a problem if the gene pool is reduced in animals that are part of a breeding programme?

Lesson outcome

You should know that waste reduction and conservation programmes are measures that can reduce or counteract the impact of pollutants on ecosystems.

4.12 More ways of reducing the effects of pollution

Link

Lesson 1.26 has more information on renewable energy sources.

Renewable resources

Renewable resources are natural sources of energy that do not run out because they can be replaced. They can be used to produce electricity and include:

- geothermal energy
- wind
- solar power
- tidal power
- biomass.

A geothermal power station in Iceland. The Earth's natural heat is used to generate electricity.

Activity A

Work in small groups. Choose one of the renewable energy sources above and find out more about it. For example:

(a) which countries use it

(b) how much of those countries' energy does it provide

(c) are there any disadvantages?

Give a 2 minute presentation to the rest of your class, outlining what you have found out about your chosen energy source.

Plant pest control. The researcher is testing insect-killing nematode worms on different citrus plants to see how effective the nematode will be at killing root weevil larvae.

Organic fertilisers and biological pest control

The effects of pollution may be reduced by using the following alternatives to *chemical* fertilisers and pesticides.

Organic fertilisers include farm manure and sewage slurry. They usually contain a lower concentration of minerals than inorganic fertiliser. Using organic fertilisers recycles animal and human waste. However, they may contain pathogens and can still cause eutrophication if they leach from fields into streams.

Biological pest control involves introducing natural enemies, such as predatory insects and parasitic wasps and **nematode worms**, to control pests.

A parasitic fly introduced into the US from Australia has successfully controlled the cottony cushion scale insect that attacks citrus fruit trees. However, there have been disastrous episodes, such as introducing the cane toad to Australia to control cane beetles. It ate all sorts of beetles, thrived and outcompeted local frogs and toads, so it has become a pest. It is poisonous and nothing in Australia can eat it.

Case study

Ravi works for a horticultural organisation. He travels around to gardening shows throughout the UK and advises people about biological control agents. Many of these can be ordered from the organisation, but the best time to use these control agents is from April to September. In some cases the biological control agent is used as part of an integrated pest management programme. The pest population is first reduced using an insecticide and then the control agent is introduced to keep the pest population very low.

1 Think back to lesson 4.3 on interdependence. How will a reduction in the pest population affect the population size of the control agent?

2 How will that, in turn, affect the size of the pest population?

The bog garden at the Royal Horticultural Society garden, Rosemoor, in Devon.

Assessment activity 4.6

| 2B.P6 | 2B.M4 | 2B.D3

You are an environmental scientist working for your local council. You have been asked by your manager to come up with ways of persuading people in your local area to recycle more.

1 Make a leaflet showing how recycling and reusing materials can reduce the impact that human activities have on the environment.

2 Write a report to describe the different methods used to help reduce the impact of human activities on ecosystems.

3 Discuss the advantages and disadvantages of the methods you have described in part 2.

4 Evaluate the success of methods to reduce the impact of human activity on an ecosystem.

Tips

For 2B.P6, give one example from your own experience of reuse and recycling.

For 2B.M4 think about what other examples you can give from your own experience of how to reuse or recycle.

For 2B.D3 choose a scenario (one example of a method used to reduce pollution) and research to find out how successful it has been, or what problems it has caused.

Lesson outcome

You should know that using renewable energy sources, organic fertilisers and biological pest control methods can reduce or counteract the impact of pollutants on ecosystems.

4.13 Infectious diseases and vaccination

Key terms

Vaccination – Procedure to inject a less harmful version of an infecting agent (microorganism) to make the body produce an immune response and make memory cells.

Pathogen – Organism that can infect another living organism and cause a disease.

Link

Refer back to lesson 4.4 to remind yourself of the characteristics of prokaryotes (bacteria).

Infectious diseases are caused by infecting agents or **pathogens**, such as bacteria and viruses.

Bacteria are single-celled **microorganisms**. They have a different cell structure from you – they are **prokaryotes**. They have a cell wall, cell membrane and cytoplasm, but no nucleus or **membrane-bound organelles**, such as mitochondria.

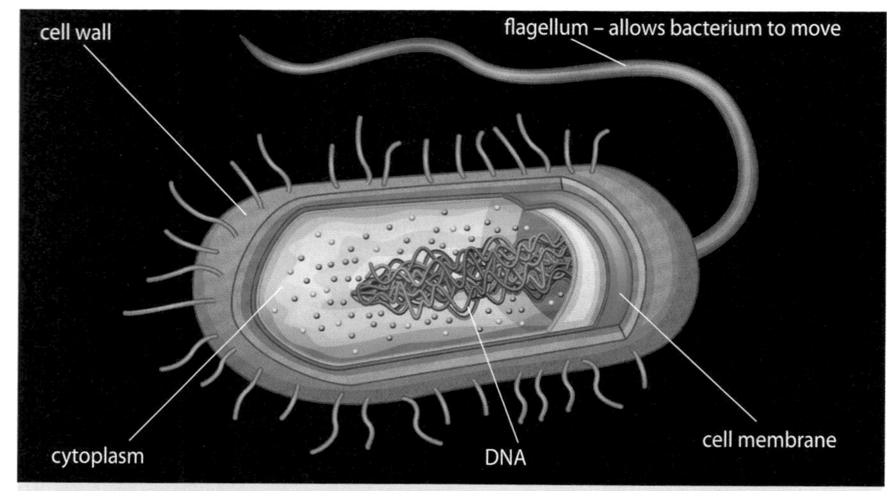

The structure of the *Salmonella* bacterium.

Bacteria can enter your body and produce toxins (poisons) that harm your cells.

Viruses are much smaller than bacteria and are not made of cells. They consist of genetic material enclosed by a protein coat. Viruses cannot reproduce on their own, so they invade your cells and use the organelles in your cells to make new virus particles. These new virus particles then break out of the infected cell, destroying it, and infect many other cells.

Viruses and bacteria can enter your body in many ways including through your skin, airways or stomach.

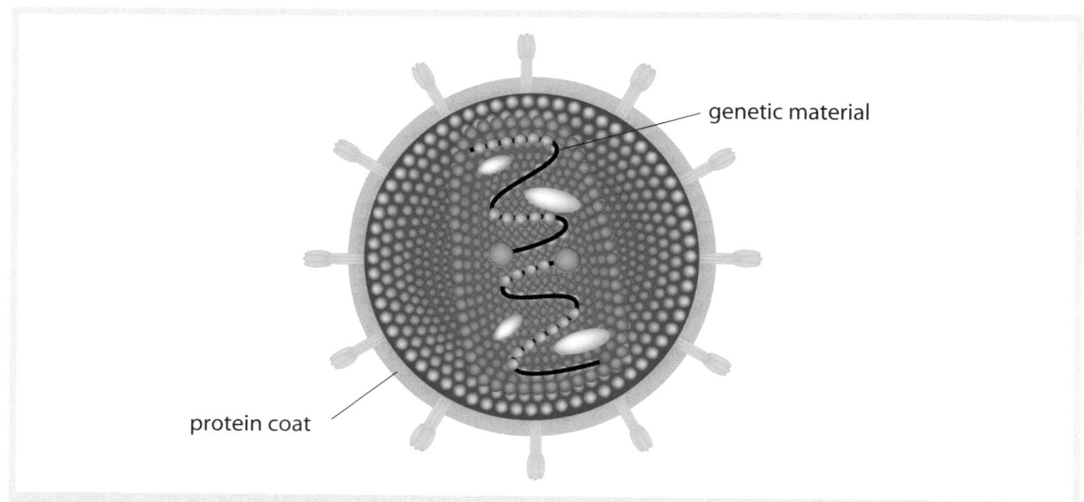

The structure of the human immunodeficiency virus (HIV). The protein coat, shown in orange, has been cut away to show the genetic material inside.

Vaccination

Vaccination can be used to prevent some infectious diseases. Normally, if a microorganism invades your body and makes you ill, your **immune system** responds by producing **antibodies** from a certain type of white blood cell.

- These white blood cells also make specific memory cells that stay in your blood for a long time.
- If the same type of microorganism gets inside your body again, these memory cells recognise it and produce lots of antibodies very quickly.
- These antibodies destroy the microorganisms before they can make you ill again. You are **immune**.

Vaccination can make you immune without you having to suffer the infection in the first place. Scientists make a vaccine that contains some dead or weakened microorganisms. You are injected with the vaccine and your body responds by making memory cells. Then, if later on you are infected by a real live microorganism of this type, you can make antibodies very quickly and not be ill.

Name of disease	Causative (infecting) agent	When vaccine is given
Measles	Virus	Babies and infants, and at 13–18 years
Mumps	Virus	Children aged 5 years
Rubella	Virus	Children aged 5 years
Whooping cough	Bacteria	Babies and infants
Polio	Virus	Babies and infants
Diphtheria	Bacteria	Babies and infants
Tetanus	Bacteria	Babies and infants
Influenza	Virus	Over 65 years or younger if at risk, e.g. health worker or asthmatic
Cervical cancer	Virus	Girls aged 12–13 years

Activity A

Make a large flow diagram to show how vaccinations work. You can research using biology textbooks and the Internet to help you.

Just checking

1 Make a list of all the ways in which bacteria are **(a)** similar to viruses and **(b)** different from viruses.

Did you know?

The white blood cells that make antibodies are called B-lymphocytes.

Discussion point

Some people choose not to have their children vaccinated against certain diseases. Why do you think this is? How might not having a child vaccinated be harmful to that child and to other children in the population?

Take it further

1 AIDS is caused by a virus, the human immunodeficiency virus (HIV). Research and find out why scientists have not yet been able to make a vaccine against this virus.

2 Research and find out about Edward Jenner's work on vaccination in the eighteenth century.

Lesson outcome

You should understand that vaccination can be used to prevent some diseases caused by microorganisms.

4.14 Antibiotics

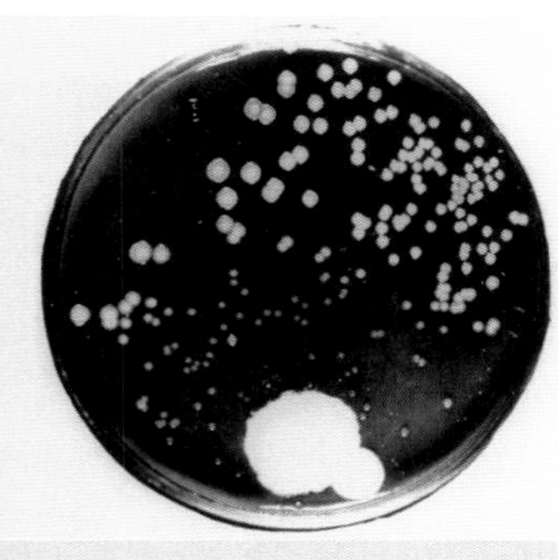

A culture plate made by Alexander Fleming. The colonies of bacteria are fewer and smaller near to the fungus, *Penicillium notatum* (in the 6 o'clock position).

Antibiotics are powerful medicines that kill or prevent the growth of bacteria and some fungi. They are produced by some microorganisms, which make them to reduce competition from other microorganisms by killing them or preventing their growth.

How were antibiotics discovered?

The first antibiotic discovered was penicillin. In 1928, the scientist Alexander Fleming came across *Penicillium* (a mould fungus) during an experiment. He realised that it stopped bacteria growing on an agar plate. By the 1940s, two other scientists, Florey and Chain, had developed the use of this antibiotic. It has since saved many lives. In 1945, Fleming, Florey and Chain shared the Nobel Prize in Physiology or Medicine for their work on penicillin. During the last 70 years, many other antibiotics have been discovered or developed by modifying existing ones.

Some examples are:

- streptomycin
- tetracycline
- gentamicin
- clarithromycin.

Did you know?

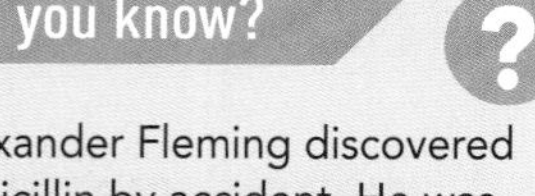

Alexander Fleming discovered penicillin by accident. He was growing bacteria on plates of nutrient medium, when one plate accidently became contaminated with spores from a fungus in a nearby lab. He almost threw away the contaminated plate, but then realised that not many bacteria were growing near the mould fungus.

How antibiotics work

Antibiotics interfere with the chemical reactions (metabolism) that occur in bacterial cells. Each antibiotic works in a slightly different way and is most effective for certain types of bacteria. Some may also produce side effects.

Doctors may first prescribe a broad-spectrum antibiotic that can deal with many types of bacteria. If that does not work, a sample, such as a throat swab, may be taken from the patient and sent to the laboratory for analysis. Once the bacterium is identified, doctors can prescribe a more specific antibiotic.

Antibiotics do not work against viruses. This is because viruses do not have cells or their own metabolism, and so there is nothing for antibiotics to interfere with.

Activity A

Some people carry a type of bacterium, called Group B *Streptococcus.* It is harmless to adults. However, it can live in the vagina of female carriers, and if it infects a newborn baby during birth is very harmful, causing a type of meningitis. Eighty babies a year die in the UK as a result of this infection, and many more become seriously ill and may suffer permanent disabilities.

There is a simple test that is carried out on pregnant women in many countries to see if they carry this bacterium. If they do, they are given penicillin during labour. Pregnant women in the UK are not routinely tested for this. They can buy a test (it currently costs about £35), but many do not know about it.

Think about the advantages and disadvantages of testing all pregnant women in the UK for this bacterium.

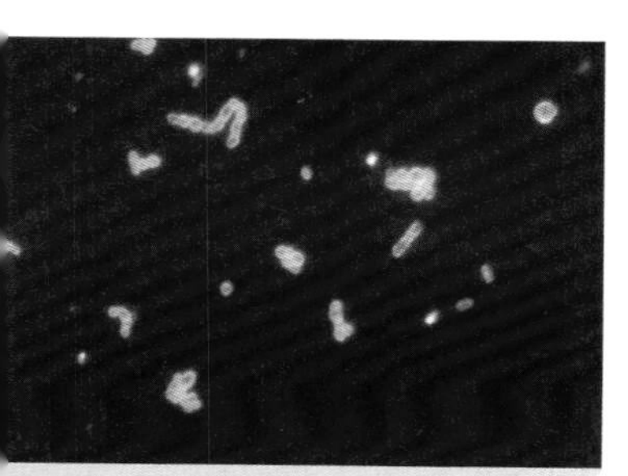

Group B *Streptococcus* bacteria.

Resistance to antibiotics

Bacteria can become **resistant** to antibiotics. This happens by natural selection. Consider this sequence of events.

- One bacterium infecting you has a mutation in its DNA.
- This makes it able to resist the effects of the antibiotic you have taken.
- It survives whereas all the other bacteria infecting you are killed by the antibiotic.
- This survivor can now flourish as it has no competition.
- It passes the gene with the mutation for **antibiotic resistance** to all its descendents, as bacteria reproduce by just dividing into two.

When you start to take antibiotics you begin to feel better as some bacteria die. But they are not all dead, and if you stop taking the antibiotics then these bacteria reproduce and you get ill again. Some of them may be resistant, so you will now need a different antibiotic.

Link

Look back to lesson 4.2 on evolution by natural selection.

Assessment activity 4.7 | 2C.P7 | 2C.M5

You are an infection control nurse in a hospital. You need to write a leaflet to inform trainee nurses, make a poster for patients and visitors and write a report to hospital management, about infecting agents such as bacteria and viruses and how bacteria become resistant to antibiotics.

1. Describe how pathogens – bacteria and viruses – affect human health.
2. Identify measures that can be taken to prevent and treat infectious diseases.
3. Explain, using some examples, how bacteria can become resistant to antibiotics.

You can present your work as one of the following: a leaflet, a poster or a report.

Tips

For 2C.P7 you need to show how bacteria and viruses act on the human body and how this affects our health.

For 2C.M5 use the NHS website and find out about hospital-acquired infections, where the bacteria are antibiotic-resistant, such as: MRSA (methicillin-resistant *Staphylococcus aureus*), C. diff (*Clostridium difficile*) and *Pseudomonas*. Show the implications this may have for the future – what happens if all bacteria become resistant to all the antibiotics we have? Show why it is important to follow the treatment regimen for antibiotics and finish the course.

Infection control in hospitals.

Just checking

1. Draw an annotated flow diagram to show how antibiotic resistance arises in a population of bacteria.
2. Explain why you should not take antibiotics unless you really need them, and why you must finish the course.

Lesson outcome

You should understand how antibiotics can be used to treat diseases caused by bacteria, and how bacteria can become resistant to antibiotics.

4.15 Lifestyle, environment and disease

Key terms

Lifestyle – Behaviours people adopt that can affect their health.

Non-infectious disease – Disease not caused by an infecting agent (pathogen).

In more developed countries the death rate from infectious diseases is low. This is due to:

- improvements in sanitation and hygiene
- better nutrition
- smaller families and better housing – people not living in overcrowded conditions
- availability of antibiotics and antiviral drugs
- introduction of vaccination programmes.

But there are many **non-infectious diseases**, which are not caused by an infecting agent and are not spread from person to person. They are treatable but may not be curable. However, they are potentially preventable.

Example of non-infectious disease	Lifestyle or environmental factor associated with it
Mental illness such as depression or schizophrenia	Misuse of recreational drugs including alcohol; also associated with misuse of some prescription drugs such as some tranquillisers
Malnutrition	The results of an unbalanced diet, eating too much or too little; some nutrients may be absent or in the wrong quantities
Deficiency disease:	Lack of certain nutrients in the diet:
rickets	lack of vitamin D
scurvy	lack of vitamin C
night blindness	lack of vitamin A
anaemia	lack of iron
Obesity	Eating too much food
Respiratory diseases including asthma, emphysema and lung cancer	Poor air quality Smoke from tobacco cigarettes
Diseases of the circulatory system, e.g. high blood pressure, heart attack and stroke	Smoke from tobacco cigarettes
Skin cancer	Over-exposure to ultraviolet light, e.g. by sunbathing without protective cream or by using sun-beds in tanning salons
Liver disease such as cirrhosis and liver cancer	Excessive consumption of alcohol

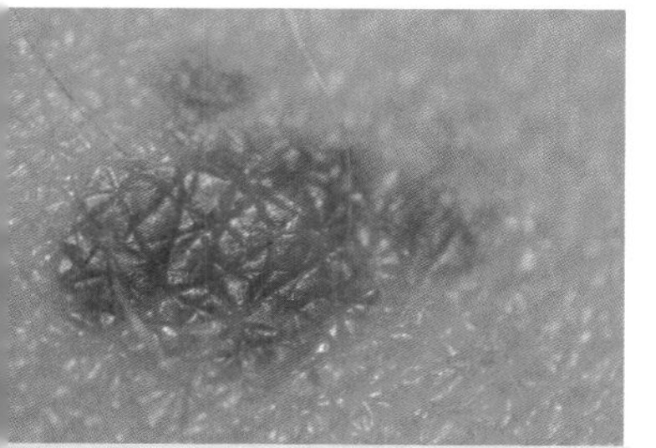

Skin cancer on a leg exposed to too much sunlight. This cancer can spread to other organs and cause death.

Did you know?

Although we are told to avoid excessive sunbathing to reduce our risk of skin cancer, some sunbathing is good for you. Sunlight stimulates our bodies to make vitamin D which protects us from other cancers and from heart disease, and makes our bones strong. However, it should be safe sunbathing, so you need to use a UV-A and UV-B filtering sun lotion.

Correlation and cause

Scientists have shown by research that lack of some nutrients causes deficiency diseases.

Some factors greatly increase your chance of getting a certain disease but they do not always cause the disease. Your genes play a part as well. So smoking tobacco, drinking too much alcohol and exposure to too much ultraviolet light are called **risk factors**.

We do not know which of us are more or less likely to develop cancer or heart disease, so it is wise to remove the **lifestyle** factor and *not* to:

- misuse drugs
- smoke tobacco
- drink too much alcohol
- sunbathe too much without protection.

Case study

Stefan works in Health Promotion. The aim of health promotion is to promote good lifestyle choices, including sensible use of alcohol, not smoking and taking plenty of exercise. This can prevent many illnesses from occurring. For every £1 spent on health promotion, the NHS saves about £9 in trying to treat those illnesses later on.

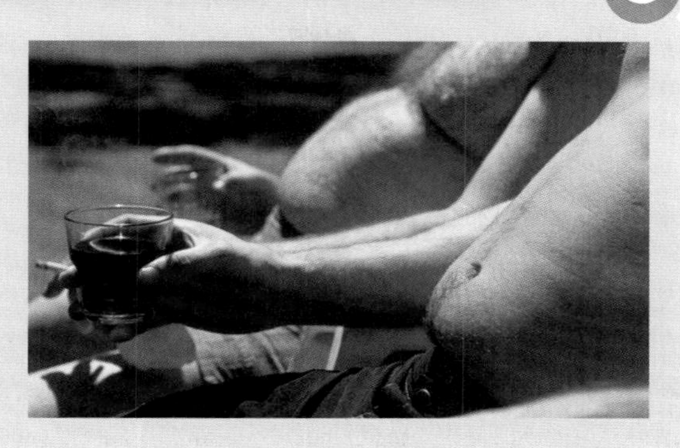

Stefan organises clinics to help people overcome their poor lifestyle choices. The idea is to educate people and empower them so they take responsibility for their health and adopt healthy behaviours. Not only is this good for the population, but it will save the NHS money that can then be spent on patients with other illnesses.

1 Why do you think that diseases like obesity, lung cancer and cirrhosis of the liver are often described as 'self-inflicted'?
2 The NHS has a large but limited budget. Some people think that people suffering from self-inflicted illnesses should pay for their treatment rather than obtain it free from the NHS. What do you think?

Adopting heathy behaviours.

Activity A

1 Use books and the Internet to research and find out the lifestyle factors that can lead to heart disease and stroke.
2 In small groups, discuss what you have found.
3 Write a list of ways to reduce the risk of getting cancer.

Assessment activity 4.8

| 2C.P9

You work in the health promotion department at a GP surgery. Design and make a poster, to display to all the patients, which describes how lifestyle choices can affect human health.

Tips

Make sure your poster is informative, but keep the number of words to a minimum. Patients must be able to read it all from a distance and in a fairly short time.

You may need to do some research using the Internet or health magazines to find the information you need.

You need to show positive effects of lifestyle choices, such as not smoking, eating healthily, not drinking too much alcohol and exercising. Also show negative effects due to smoking, recreational drug use (including alcohol), having a poor diet and not exercising.

Lesson outcome

You should understand how non-infectious diseases can be caused by lifestyle and environment.

4.16 Inherited diseases

Some genetic diseases can be inherited. This means they can pass from parents to offspring. Sometimes the diseases skip a generation and pass from grandparents to grandchildren. The members of the generation in between *carry* the alleles without having the disease.

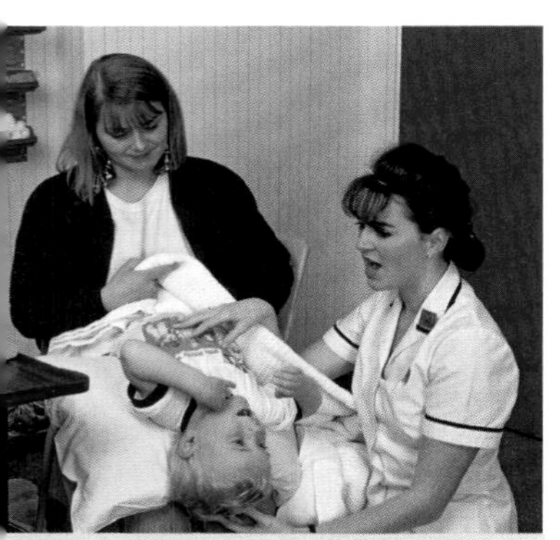

Patient with cystic fibrosis having physiotherapy to remove the excess, thick mucus from his lungs.

Cystic fibrosis

Cystic fibrosis is the most common inherited disease in the UK. It has a recessive inheritance pattern. Both parents may be carriers without having any symptoms. If both egg and sperm have a faulty allele, the resulting baby will have cystic fibrosis. The main symptom is thick sticky mucus in the lungs and gut. This leads to chest infections and poor growth.

Sufferers are treated with enzymes to help digest food, physiotherapy to remove the mucus from their lungs and antibiotics to prevent lung infections. If their lungs become too damaged they may later need a heart–lung transplant.

Huntington disease

Huntington disease is a much rarer genetic disease. It affects the nervous system and gradually leads to lack of control over movements and to dementia. Sufferers do not usually have symptoms until they are aged over 40 years.

Huntington disease has a dominant inheritance pattern, which means that even if only one parent is affected the children have a 50:50 chance of inheriting it.

Remember

You may remember from lessons 1.5 and 1.6 that some mutations in genes produce alleles that can give rise to genetic variations that can be inherited.

You have used genetic diagrams, pedigree analysis and Punnett squares to show how characteristics can be inherited.

Did you know?

Although people with Huntington disease do not normally have symptoms until they are over 40, a boy in South America, from a family where many members had Huntington disease, developed symptoms at the age of 2 years.

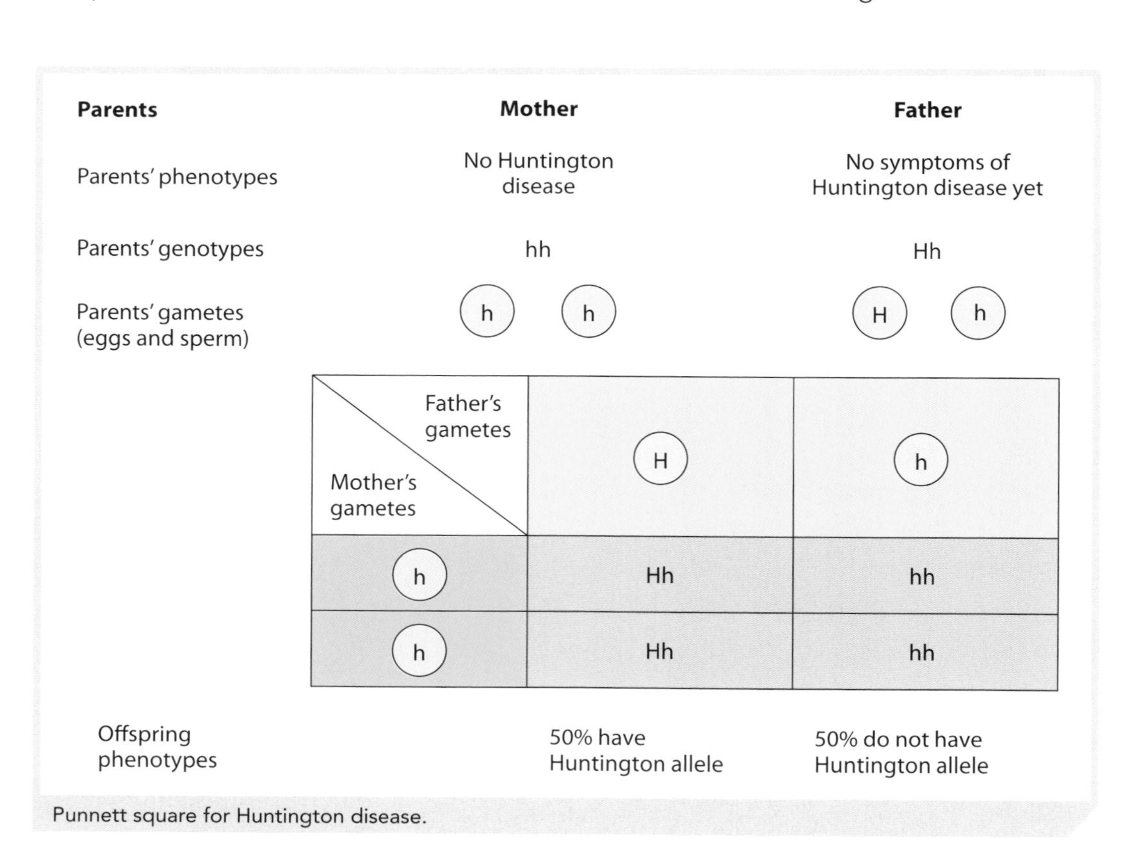

Parents	Mother	Father
Parents' phenotypes	No Huntington disease	No symptoms of Huntington disease yet
Parents' genotypes	hh	Hh
Parents' gametes (eggs and sperm)	h h	H h

Father's gametes / Mother's gametes	H	h
h	Hh	hh
h	Hh	hh

Offspring phenotypes	50% have Huntington allele	50% do not have Huntington allele

Punnett square for Huntington disease.

Activity A

Look at the Punnett square on the previous page and the family tree on the right. Jack has had a genetic test and knows that he has an allele for Huntington disease. However, he is only 25 years old and has not developed any symptoms yet.

If Jack marries someone who knows that she does not have the allele for Huntington disease, what are the chances of them having a child with the Huntington allele?

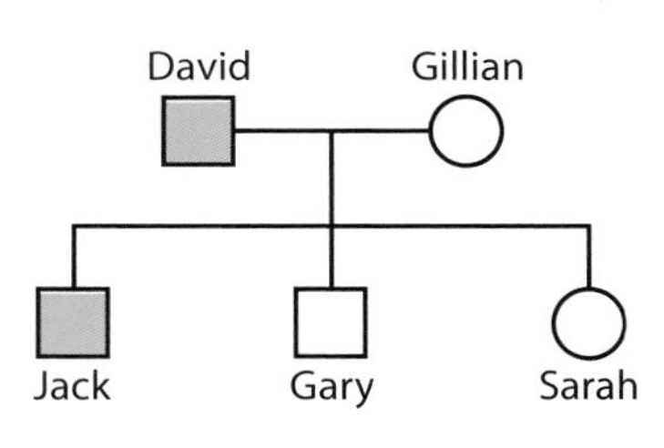

Key

- male with Huntington allele
- unaffected male
- unaffected female

Family tree showing the chances of inheriting Huntington disease.

Assessment activity 4.9

| 2C.P8 | 2C.M6 | 2C.M7 | 2C.D4

You are a journalist writing an article for the health pages of a magazine. Some people are very confused about the causes of ill health and do not understand that their lifestyle can play a part. Some people think that vaccinations cure diseases, and that antibiotics can be used for all sorts of illnesses, including those caused by viruses. Many people have heard of inherited diseases, but do not understand how these are passed from parents to children. Your article should inform people about factors that can affect human health. Focus on the differences between prevention, treatment and cure, and how people can use pedigree analysis to work out their chances of having a baby with an inherited disorder.

Your article should include the following information.

1. A list of the different biological, social and inherited factors that affect human health.
2. A description of two different treatment regimes – one used to prevent a disease and one used to treat a disease.
3. An explanation of the use of pedigree analysis.
4. A discussion of the advantages and disadvantages of vaccination programmes.
5. An evaluation (benefits and risks) of the use of antibiotics, pedigree analysis and vaccination programmes in the treatment and prevention of childhood diseases.

Tips

For 2C.P8 remember that vaccinations prevent diseases – they do not treat them. Antibiotics treat some diseases – those caused by bacteria and fungi.

For 2C.M6 use at least one example of pedigree analysis to explain how pedigree analysis is used.

For 2C.M7 think about the following. Concerns about the MMR vaccine meant many parents did not get their children vaccinated against measles. Outline the disadvantages of children not being vaccinated and weigh these against the very small risk of vaccines harming some children.

For 2C.D4 you need to find some statistics and data – for example how the incidence of a disease has reduced after a vaccination programme has been introduced. Include some reference to how many lives have been saved by antibiotics. Think about the value of pedigree analysis now that there are genetic tests available to see if people carry a faulty allele.

Take it further

Use textbooks and the Internet to find out about familial high blood cholesterol and how it is inherited. In small groups, make a poster showing what you have found. Share your ideas with the rest of the class.

Lesson outcome

You should understand that some diseases are inherited, and pedigree analysis can be used to show the inheritance of genetic disease.

4.17 Physical activity keeps the body healthy

You may have noticed certain changes in your body when you exercise:

- your heart rate increases
- your rate and depth of breathing increases
- you feel hot and sweat more
- your skin flushes red.

Why do these changes happen?

Link

Look back at lesson 1.2 to remind yourself about the different types of muscle cells; look at lesson 1.3 to revise organs and organ systems; read lesson 1.9 to remember how your skin helps regulate your body temperature.

- When you undertake physical activity, your muscles contract more. For this they need energy, so the rate of respiration in the muscle tissue increases.
- At first, not enough oxygen is supplied to the muscles to increase **aerobic respiration** enough to meet this extra demand. So your muscles also use **anaerobic respiration** to release energy.
- However, within a few minutes your heart muscle contracts more forcefully and more blood is pumped out at each beat. Your heart rate also goes up.
- Together, this means more blood is pumped out from your heart each minute. This blood flows in arteries to your muscles and delivers more oxygen for aerobic respiration.
- As your muscle tissue continues to respire, it produces more carbon dioxide and more heat. Your rate and depth of breathing increase to remove the extra carbon dioxide. This increase in breathing will also get more air, containing oxygen, into your lungs.
- Blood carries the extra heat away from muscles. As your blood flows through your skin, it loses its heat by vasodilation and via evaporation of water in sweat.

Cycling uses a lot of oxygen and increases your rate of aerobic respiration. Your muscles respire fat.

If you take physical exercise often and regularly then these changes happening in your body help to keep you healthy. It can improve your cardiovascular system, respiratory system, musculoskeletal system and immune system. It also has great psychological benefits and makes you feel better.

System	How exercise improves it
Cardiovascular system	• Your heart muscle becomes stronger and can beat more powerfully. • There is less risk of fatty deposits forming in your artery walls. • Your resting blood pressure is reduced.
Respiratory system	• Your depth of breathing increases. • Your lungs can pass more oxygen into the blood.
Musculoskeletal system	• Your muscles get bigger and more powerful. • Your muscle fibres produce more mitochondria and enzymes for respiration. • Your bone density increases so your bones are stronger. • Your tendons and ligaments become stronger. • Your backbone becomes stronger and your joints are more flexible.

Activity A

You are a personal fitness trainer. Make a large poster to show your clients the benefits of exercise on health. Include a diagram of the human body and annotate it to indicate the various systems and how exercise affects them.

How exercise improves your general health

- The decrease in resting blood pressure and decrease in fatty deposits in the walls of arteries help to reduce your risk of heart attack and stroke.
- When muscles work hard they respire fats. This helps you to maintain a healthy weight.
- Your balance, coordination, flexibility, stamina and strength improve.
- You have less chance of lower back pain.
- Having stronger bones reduces your risk of breaking bones and reduces your risk of **osteoporosis** in later life.
- Your immune system will work better so you are more resistant to infectious diseases.
- When you exercise, your brain makes chemicals called endorphins. These make you feel good, so exercise makes you happier. This is an important aspect of your mental health.
- Physical exercise also helps you concentrate on intellectual tasks. It improves your brainpower and your ability to learn.

Did you know?

You reach your maximum bone density in your early twenties. It is important to exercise when young, so that you reach a good maximum bone density. It is also important to keep doing exercise as an adult, so you will be less likely to suffer from osteoporosis later in life.

Weight-bearing activities such as running, jumping, skipping, dancing and walking are all good for increasing bone density.

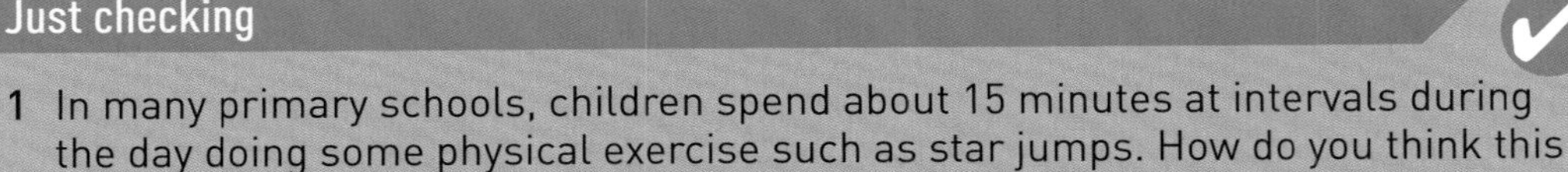

Just checking

1 In many primary schools, children spend about 15 minutes at intervals during the day doing some physical exercise such as star jumps. How do you think this helps their learning (e.g. reading, writing, maths) skills?

2 Scientists have found that people who do some physical exercise each day, suffer from fewer colds than people who don't exercise. Why do you think this is?

Lesson outcome

You should understand how physical activity helps to keep the body healthy.

Research and investigation skills

Key terms

Independent variable – A variable that is deliberately changed during the investigation.

Dependent variable – A variable which may change as a result of changes to the independent variable and is measured during the investigation.

Control variable – A variable which must be kept constant during the investigation if it is to be a fair test.

What is an investigation?

In an investigation, engineers, technicians and scientists aim to answer a technical or scientific question. Often they will make a prediction about the answer to the question and then test their prediction using observations or measurements.

In the workplace this could involve:

- a building technician investigating different types of insulation for cavity walls
- an aerodynamicist investigating which design of rear wing gives the greatest downforce on a Formula 1 car.

The stages of an investigation

As part of their investigations, engineers and scientists carry out some or all of the following stages. You will do the same when you carry out investigations as part of this course.

- Decide on a suitable question.
- Carry out some reading and research.
- Use this research to produce a prediction, called a **hypothesis**, that you can test.
- Use this research to plan the practical work that you will do to test this prediction.
- Make a risk assessment for the practical work.
- Carry out the practical work and make observations or measurements.
- Analyse the evidence from these observations and measurements to reach a conclusion and test the prediction.

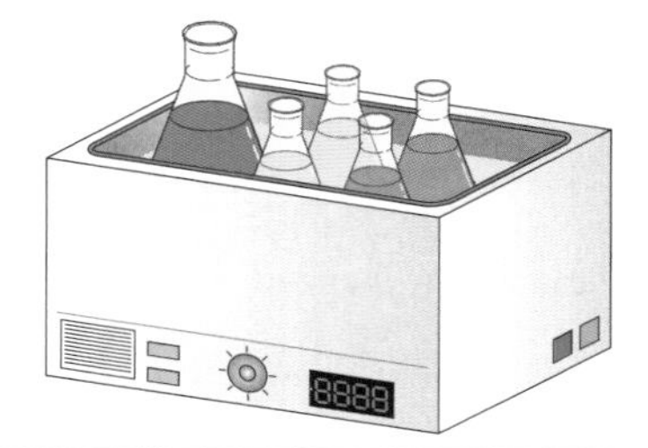

A water bath is a good way of making sure that the temperature of a reaction remains the same during an investigation.

Controlling variables

Things that you change or measure in an investigation are called variables. For an investigation to give valid results, it must be a fair test of how the **independent variable** affects the **dependent variable**. It is very important that you make sure no other factors change during the investigation. These other factors are called **control variables**.

Activity A

Adam works as an engineering technician. He wants to find out how the current in a resistor made out of a new material depends on the voltage applied to it. He changes the voltage in the circuit between 2 V and 10 V and measures the current in the resistor at a range of voltages. He makes sure that he uses the same resistor in each experiment.

Name the independent variable, the dependent variable and the control variable.

Making measurements

When you are planning an investigation you need to ask yourself:

- What do I need to measure or observe?
- What shall I use to take measurements?
- How many measurements do I need to take, or how often?

Case study

Lucy works as an environmental technician investigating the pollution in lakes, canals and rivers. What decisions does she need to make about measurements? Look back to Lesson 4.10 and see what she needs to measure. Then decide what equipment she will use and devise a sampling method for her.

Risk assessments

By law, anyone carrying out a hazardous procedure must carry out a **risk assessment**. This is an examination of the **hazards** in an activity and the steps taken to reduce the **risk** of these hazards. Practical work in laboratories may be hazardous so your teacher must ensure a risk assessment has been carried out before you start any practical work.

Hazards could be connected with particular chemicals, but could also include:

- handling radioactive substances
- handling very hot apparatus
- disease risks from handling wildlife
- the possibility of accidents while collecting samples from the environment.

Substance being used	Nature of hazard	Steps taken to minimise risk
Sodium hydroxide solution	Irritant–particularly to the eyes	Wear eye protection Wipe up spills immediately Wash off skin immediately

This is a good example of how to set out a risk assessment.

General laboratory safety

To reduce all risks you need to make sure that you are following common-sense rules about laboratory safety.

- Move around the laboratory carefully and slowly.
- Never taste a chemical unless you are told to do so by your teacher.
- Do not eat and drink in the laboratory.
- Tie up loose hair, especially when heating substances.
- Wear eye protection.

Displaying results and interpreting data

Key terms

Anomalous result – A result that does not appear to fit with the pattern of other results in an experiment.

Error – Something which happens during an experiment which causes results to be different to the true value.

In any job it is important to be able to communicate clearly with your colleagues. In scientific and technical jobs this will often mean being able to display your data in a way that others can understand and being able to describe exactly what you have discovered by carrying out experimental work.

Tables of results

In most investigations you will need to put your results into a table. You also need to think about the best way of laying out your table and include appropriate headings and units as shown in this example. The data are arranged in ascending order of the independent variable.

The measurements of the variables are organised into columns.

There are helpful labels at the top of each column, including correct units.

Voltage (V)	Current (A)
2.0	0.24
4.0	0.46
6.0	0.70
8.0	0.82
10.0	1.20

The first column is the independent variable – what the experimenter changes.

The second column is the dependent variable – what the experimenter measures.

Identifying patterns: if your table is set out like this, you can often see a pattern in the data.

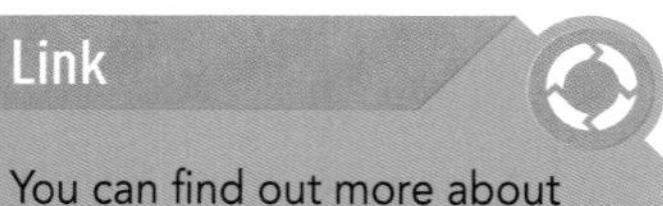

Link

You can find out more about this investigation in Activity A on page 196.

Activity A

For the results in the table above, describe what happens to the current when the voltage is increased.

Remember, in an investigation you are often trying to find out how one variable depends on another – in this case what happens to the current when the voltage is increased.

Take it further

When a line of best-fit is a straight line going through the origin (where both variables are zero), this means the two variables are proportional to each other.

If voltage is doubled then the current will double as well.

Using graphs

You can often see the pattern in data just by looking at a well-designed table. But it is always a good idea to go on to display the data in a graph because it:

- makes the patterns much easier to see
- helps you to spot anomalous results
- can help you to write mathematical relationships between variables
- helps you to predict values of variables for which you have no data.

Line graphs

When data are **continuous**, you should plot a line graph. The independent variable should be plotted on the horizontal axis and the dependent variable on the vertical axis. Then you can draw a line of best-fit passing as close to the data points as possible.

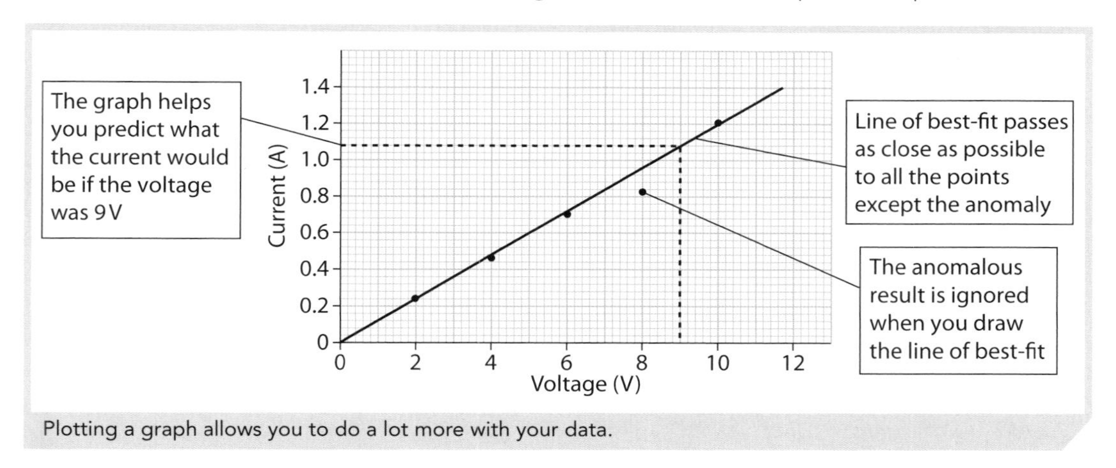

Plotting a graph allows you to do a lot more with your data.

Bar charts

You can use bar charts when the independent variable is not continuous, or does not have numerical values.

Lucy, an environmental technician, collected the data below for oxygen concentrations for water taken from different sources. The bar chart shows clearly how much lower the concentration of oxygen is in the canal water.

Water source	Oxygen concentration (mg/dm^3)
Lake	5.2
River	6
Canal	3.1

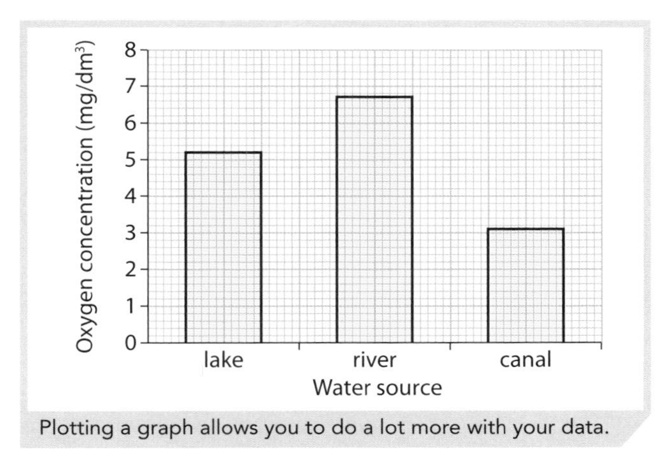

Plotting a graph allows you to do a lot more with your data.

Anomalous results

No experiment will ever be perfect. You will often find that there are **anomalous results** in your data.

If you plot your results on a line graph, you can easily see which results are anomalous. If possible you should reinvestigate them. They should not be used in calculating any averages or for plotting a line of best-fit.

Did you know?

Anomalous results play an important role in scientific research. If results are anomalous and cannot be explained by experimental error then it can lead scientists to propose new theories to explain them.

Activity B

Look at the graph of voltage plotted against current. Which result was anomalous? How can you tell?

Errors

Errors can occur for different reasons.

Eleanor is investigating how the rate of a reaction between hydrochloric acid and calcium carbonate depends on the size of the calcium carbonate pieces. She collects the carbon dioxide in a gas syringe and measures the volume of gas produced at different times.

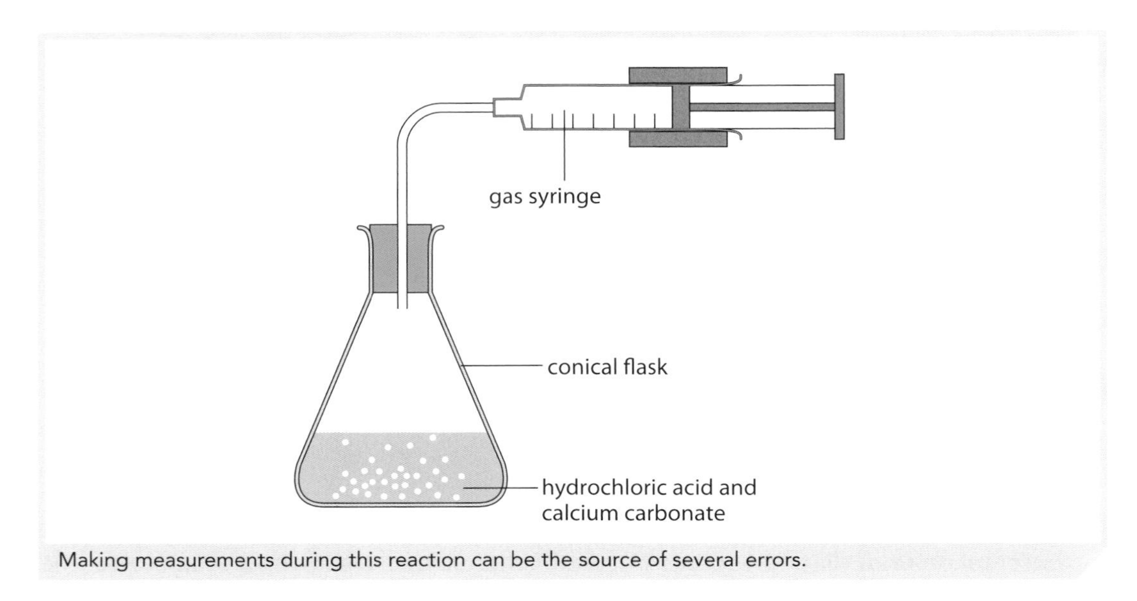

Making measurements during this reaction can be the source of several errors.

There are several reasons for errors:

- The precision of the measuring instrument. Eleanor is using a gas syringe which measures volumes to the nearest cm^3. If the gas syringe only had markings every 5 cm^3 then the measurement she wrote down might not be the true value.
- The skill of the experimenter. Eleanor may need to read the stop clock and the gas syringe at the same time so there will be an error in either the reading of the time or of the volume.

Valid conclusions

You will sometimes be asked to comment about whether a conclusion is valid. There are two things that you should think about.

Is the conclusion based on a fair test? If not, then the conclusion is probably not valid.

Are the errors significant? If there are a large number of anomalous results and the errors in carrying out the experiment and making measurements are very large then the conclusion may not be valid.

The periodic table of elements

The periodic table of elements

Key

relative atomic mass **atomic symbol** name atomic (proton) number

1	2											3	4	5	6	7	0
							1 **H** hydrogen 1										4 **He** helium 2
7 **Li** lithium 3	9 **Be** beryllium 4											11 **B** boron 5	12 **C** carbon 6	14 **N** nitrogen 7	16 **O** oxygen 8	19 **F** fluorine 9	20 **Ne** neon 10
23 **Na** sodium 11	24 **Mg** magnesium 12											27 **Al** aluminium 13	28 **Si** silicon 14	31 **P** phosphorous 15	32 **S** sulfur 16	35.5 **Cl** chlorine 17	40 **Ar** argon 18
39 **K** potassium 19	40 **Ca** calcium 20	45 **Sc** scandium 21	48 **Ti** titanium 22	51 **V** vanadium 23	52 **Cr** chromium 24	55 **Mn** manganese 25	56 **Fe** iron 26	59 **Co** cobalt 27	59 **Ni** nickel 28	63.5 **Cu** copper 29	65 **Zn** zinc 30	70 **Ga** gallium 31	73 **Ge** germanium 32	75 **As** arsenic 33	79 **Se** selenium 34	80 **Br** bromine 35	84 **Kr** krypton 36
85 **Rb** rubidium 37	88 **Sr** strontium 38	89 **Y** yttrium 39	91 **Zr** zirconium 40	93 **Nb** niobium 41	96 **Mo** molybdenum 42	[98] **Tc** technetium 43	101 **Ru** ruthenium 44	103 **Rh** rhodium 45	106 **Pd** palladium 46	108 **Ag** silver 47	112 **Cd** cadmium 48	115 **In** indium 49	119 **Sn** tin 50	122 **Sb** antimony 51	128 **Te** tellurium 52	127 **I** iodine 53	131 **Xe** xenon 54
133 **Cs** caesium 55	137 **Ba** barium 56	139 **La*** lanthanum 57	178 **Hf** hafnium 72	181 **Ta** tantalum 73	184 **W** tungsten 74	186 **Re** rhenium 75	190 **Os** osmium 76	192 **Ir** iridium 77	195 **Pt** platinum 78	197 **Au** gold 79	201 **Hg** mercury 80	204 **Tl** thallium 81	207 **Pb** lead 82	209 **Bi** bismuth 83	[209] **Po** polonium 84	[210] **At** astatine 85	[222] **Rn** radon 86
[223] **Fr** francium 87	[226] **Ra** radium 88	[227] **Ac*** actinium 89															

** The lanthanoids (atomic numbers 58–71) and the actinoids (atomic numbers 90-103) have been omitted.*

The relative atomic masses of copper and chlorine have not been rounded to the nearest whole number.
For radioactive isotopes the relative atomic mass number is given in square brackets.

Glossary

acid – A solution with a pH of less than 7.
acid rain – Rain that is more acidic than normal, due to sulfur dioxide and nitrogen oxides dissolved in it.
adaptations – The differences in structure that allow an organelle/cell/tissue/organ/organism to carry out its functions.
adenine – One of the four bases in DNA. It forms a base pair with thymine.
aerobic respiration – The process, in cells, of releasing energy that requires oxygen.
algal bloom – Rapid and excessive growth of algae, usually caused by high nutrient levels (and other favourable conditions).
alkali – A base that is soluble in water. Alkalis have a pH of greater than 7.
allele – A version of a gene.
alternating current – A current that regularly changes direction with time.
ammeter – Instrument that measures current.
amplitude – The maximum displacement of a wave from its equilibrium position.
anaerobic respiration – The process, in cells, of releasing energy in the absence of oxygen.
anomalous result – A result that does not appear to fit with the pattern of other results in an experiment.
antibiotic – Substance that can kill or slow the growth of bacteria or fungi.
antibiotic resistance – Bacteria or fungi that have the ability to survive/are not killed in the presence of antibiotics.
antibodies – Proteins produced by a type of white blood cell that attack and kill pathogens such as bacteria or viruses.
artery – A vessel carrying blood away from the heart (usually oxygenated blood).
asteroid – A small, rocky object that orbits the Sun.
atom – The smallest part of an element that can take part in a chemical reaction.
atom economy – A measurement of how efficient an industrial process is; a reaction has a high atom economy if most of the atoms in the reactants end up in useful products.
atomic number – The number of protons in the nucleus of an atom. Also known as the proton number.

bacteria – Single-celled microorganisms without a nucleus; one of the kingdoms of organisms.
base – A substance that can neutralise an acid. Bases are often metal oxides or metal hydroxides.
battery – Two or more electrical cells working together.
becquerels (Bq) – The unit of radioactivity.
Big Bang theory – The theory that the Universe began with an explosion.
bioaccumulation – The accumulation of a substance in the tissues of living organisms.
biodiesel – A biofuel made from plant oils. It can be used as a fuel in diesel cars.
bioethanol – Ethanol produced from plants, e.g. sugar or corn, that can be used as an alternative fuel for transport.
biogas – A biofuel made by fermenting manure or other kinds of waste. It can be used for heating or to produce electricity.
bonding – The way in which particles are held together in chemical substances.

cancer – A group of diseases caused by uncontrolled cell division leading to the formation of a tumour or a growth.
cardiovascular system – The heart, arteries, veins and blood.
carnivore – An animal or plant that feeds on animals.
catalyst – A substance that speeds up a chemical reaction but is not itself used up, so it can be used again.
cell (battery) – A device used to supply electrical energy from chemical energy. An electrical cell is a single unit containing two electrodes separated by an electrolyte.
cell (biological) – The basic unit of all known living organisms.
cellulose – A polysaccharide that makes up plant cell walls.
central nervous system (CNS) – The brain and spinal cord.
chain reaction – A reaction that keeps itself going once it has started.
chemical energy – Energy stored in the chemical bonds of substances.
chemical property – Something we can observe about a substance that involves a chemical change; for example, how it reacts with water.
chlorophyll – Green pigment required for photosynthesis; found in the chloroplasts of plant cells and in some bacteria and algae.
chloroplast – An organelle in plant cells. It contains chlorophyll and is where photosynthesis takes place.
chromosome – A long thread of a molecule called DNA. Each chromosome contains a series of genes along its length.
climate change – Changes to the Earth's climate, or changes in weather patterns on a global scale.
combustion – Chemical reaction when substances burn, combining with oxygen to produce heat and products such as carbon dioxide and water.
comet – Object made of frozen gas and ice that has come from outside our Solar System.
complementary base pairing – The pairing of adenine with thymine and cytosine with guanine in the DNA double helix.
compound – A substance made up of two or more different elements chemically bonded together.
concentration – The amount of a solute dissolved in a certain volume of solution. It can be measured in the units g/dm^3 (grams per cubic decimetre).
conduction – The transfer of thermal energy because of a temperature difference between one end of a material and the other.
conductor – A material that can conduct heat and electricity easily. Metals are good conductors.
conservation of energy – This is the principle that energy cannot be created or destroyed. It can only transform from one form to another.
consumer – An organism that obtains its energy by eating other organisms.
Contact process – The industrial process which produces sulfuric acid from sulfur.
continuous data – Data that can take any value within a range, not just certain fixed values.
control rods – These are rods, usually made of boron, that absorb neutrons to control the speed of the reaction or stop the fission reaction altogether.
control variable – A variable which must be kept constant during an investigation if it is to be a fair test.
convection – The transfer of thermal energy by the movement of a liquid or gas.
cosmic microwave background radiation – Electromagnetic energy that comes from all directions in space, and is believed to have come from the Big Bang.
covalent bonding – Bonding between atoms that are sharing one or more pairs of electrons.
current – The flow of charge in an electrical circuit.
cytosine – One of the four bases in DNA. It forms a base pair with guanine.

daughter nucleus – A nucleus produced when the nucleus of an unstable atom splits into two during fission or when a radioactive nucleus decays by emitting an alpha or beta particle.
decay (radioactive) – When an unstable nucleus changes by giving out ionising radiation to become more stable.
decomposer – Organism that breaks down dead material, causing decay.
deforestation – The clearing of large areas of forest to use the land for other purposes, such as growing crops, or to use the timber for fuel, building materials or making furniture.
dependent variable – A variable which may change as a result of changes to the independent variable and is measured during an investigation.
detritivore – Animal that feeds on partly broken down pieces of plant or animal tissue (detritus).
diabetes – A disease in which blood glucose concentration is not controlled. In type 1 diabetes, the pancreas doesn't make enough insulin. In type 2 diabetes, the body can't respond normally to the insulin that is made.
diabetic – A person who suffers from diabetes.
differentiation – The specialisation of cells to perform a specific job.
diffusion – Movement of molecules from a region where they are in high concentration (lots of them) to a region where they are in lower concentration.
direct current – A current that only flows in one direction.
displacement – A reaction in which a more reactive element displaces a less reactive element from a solution of its compound.

distillation – A method for separating a liquid from a mixture by evaporating and condensing it.
dominant – An allele of a gene that is always expressed in the phenotype, even if only one of these alleles is present in the cells. Represented in genetic diagrams by a capital letter.
double helix – The shape of the long DNA molecule, with two strands twisting around one another.

ecosystem – All the organisms living in a particular area and the area itself. The organisms interact with each other and with the non-living features of the area.
effector – A cell or an organ that brings about a response to a stimulus.
efficiency (energy) – The useful energy output from a power plant divided by the energy put in.
elastic potential energy – The energy stored by things that have been stretched or squashed and can spring back.
electric charge – Charged particles that transfer electrical energy in electrical devices.
electrode – The positive or negative pole of a cell or battery.
electrolyte – A liquid which is able to conduct electricity, for example in a battery or cell.
electromagnetic radiation – A form of energy transfer, including radio waves, microwaves, infrared, visible light, ultraviolet, X-rays and gamma rays.
electromagnetic wave – A wave made up of electric and magnetic fields, for example radio waves, infrared radiation, visible light and X-rays.
electron – Negatively charged subatomic particle found in shells around the nucleus of an atom. An electric current in a metal wire consists of moving electrons.
electron shells – The positions that electrons can occupy around an atom, also known as energy levels.
electronic configuration – The way electrons are arranged in the shells of an atom.
element – A substance which contains just one type of atom.
elliptical – A shape like a flattened circle.
embryo – A new individual in its first stage of development.
endocrine system – A system of glands that release hormones into the blood.
energy levels – See electron shells.
energy transfer – Energy being moved from one place to another, possibly with a change in the form of energy at the same time.
enzyme – A protein made by cells that acts as a catalyst (speeds up chemical reactions).
error – Something which happens during an experiment which causes results to be different to the true value.
eukaryotes – Organisms made of a cell or cells in which the genetic material is enclosed in a nucleus.
eutrophication – Poisoning of areas of water (streams, rivers or lakes) by excess nitrates and phosphates, which cause algal bloom, death of plants, depletion of oxygen and death of fish.
evaporation – When a liquid turns to a vapour or a gas.
evolution – Gradual change over a period of time.
excretion – The discarding of waste products from within the body, such as urine from kidneys.

fallout – Radioactive dust and ash that is carried through the air after a nuclear accident and falls to Earth far from the accident site.
fault – A crack in the Earth's crust caused by the movement of rock on either side of it.
ferment/fermentation – One type of anaerobic respiration by microorganisms.
filtration – Method used to separate an insoluble solid from a liquid.
fluid – Any substance that can flow; liquids and gases are fluids.
forensic science – The application of any branch of science to answer questions of a legal nature and aid criminal investigations.
formula (chemical) – A way of showing the number of atoms of each type which bond together.
fossil fuel – Non-renewable fuels, such as coal, oil and natural gas that have formed over millions of years from dead plants and animals.
frequency – The number of waves in one second.
fuel rods – These are rods that contain the material, for example uranium, that splits during the nuclear fission reaction.
fungi – Multicellular eukaryotes, non-photosynthetic and have cell walls containing chitin. Some are parasitic and some feed on dead or decaying matter. One of the kingdoms of organisms.

galaxy – Billions of stars that are held together by the force of gravity. They tend to move about a central mass.
gene – A section of DNA that carries the instructions for a characteristic.
generator – A machine that makes electricity when it turns.
genotype – Shows the alleles present in an individual, for a particular characteristic.
gland – A part of the body that makes useful substances and then releases them.
global warming – The increase in the Earth's average temperature likely to be caused by increased amounts of greenhouse gases in the atmosphere.
glucagon – A hormone produced in the pancreas when blood glucose levels get too low; acts on liver cells and causes stored glycogen in them to be changed to glucose and released into blood.
glycogen – Type of carbohydrate, sometimes called 'animal starch'. It is a large molecule made of many glucose molecules joined together.
gradient – The steepness of a line. Resistance is represented by the gradient of a voltage-current graph.
gravitational potential energy – The energy stored in things because of their position.
greenhouse gases – Gases that help to trap heat in the atmosphere. They include carbon dioxide, methane and water vapour.
guanine – One of the four bases in DNA. It forms a base pair with cytosine.

Haber process – Industrial process used to produce ammonia.
habitat – The place where an organism lives.
haemoglobin – The red iron-containing pigment found in red blood cells.
half-life – The time it takes for half the nuclei in a sample of a radioactive isotope to decay.
halides – Compounds formed from group 7 elements.
halogens – Elements in group 7 of the periodic table.
hazard – Something with the potential to cause harm.
herbicide – Chemical that kills plants, usually used on weeds.
herbivore – An animal that feeds on plants.
heterozygous – If the two alleles of a gene for a characteristic are different, the organism is heterozygous for that characteristic.
homeostasis – Keeping the body's internal conditions in a steady state.
homozygous – If both alleles of a gene for a characteristic are the same, the organism is homozygous for that characteristic.
hormone – A substance that is made and released in one part of the body and that has an effect on another part of the body (a chemical messenger).
hybrid – A vehicle that runs partly on an electric motor, powered by a battery, and partly on a conventional internal combustion engine.
hypothesis – A prediction that can be tested.

identification key – A tool used to identify different biological organisms.
immune – Protected from a disease by the presence of antibodies, or memory cells that can quickly make antibodies.
immune system – The organs and mechanisms that protect an organism against pathogens and disease.
independent variable – A variable that is deliberately changed during an investigation.
infrared – Electromagnetic radiation that we can feel as heat.
insulator – Material that is a poor conductor of heat and electricity. Plastics, ceramics (pottery) and air are good insulators.
insulin – A hormone produced by the pancreas that controls the concentration of glucose in the blood.
interdependence – How organisms in an ecosystem depend on each other to survive.
involuntary response – One not under conscious control; does not involve thought.
ion – An atom or group of atoms that is electrically charged.
ionic bonding – Bonding between positive and negative ions.
irregular (galaxy) – A galaxy that does not have a particular shape, unlike a spiral or an elliptical galaxy.
isotopes – Atoms of the same element which have the same number of protons but different numbers of neutrons.

joule – The joule (symbol J) is the scientific unit used to measure energy.

kinetic energy – The energy an object has because it is moving.

lava – Molten rock on the Earth's surface.
lifestyle – Behaviours people adopt that can affect their health.
light year – The distance that light travels in a year – about 9.5 thousand billion km or about 6 thousand million miles.
liver – A large organ in the body that makes bile and carries out many chemical reactions/makes many chemicals needed to sustain life; particularly important in dealing with toxins.

magma – Molten rock inside the Earth's crust.
mantle – The part of the Earth between the crust and the core.
mass number – The total number of protons and neutrons in the nucleus of an atom. Also known as the nucleon number.
mechanical energy – The sum of the kinetic energy of an object and its potential energy.
membrane-bound organelles – Organelles with membranes around them, e.g. nucleus and mitochondria found in eukaryotic cells. Prokaryote cells have organelles (e.g. ribosomes) but they do not have any membrane-bound organelles.
meteor – A piece of debris from space that glows as it enters the Earth's atmosphere (also known as a shooting star).
microorganism – An organism that can only be seen with the aid of a microscope.
Milky Way – The name of the galaxy that contains our Solar System.
mitochondrion (plural mitochondria) – The organelle where aerobic respiration takes place in eukaryotic cells.
mixture – A single substance made up of two or more simpler substances that are not chemically bonded.
molecule – A particle made up of two or more atoms bonded together.
motor neurone – A neurone that carries impulses from the central nervous system to effectors (e.g. muscles and glands).
mutation – A change to genetic material (DNA or chromosomes) that may lead to a change in a characteristic.

natural selection – Mechanism for evolution. The best-adapted organisms survive and reproduce, and the alleles for the favourable characteristics are passed on to their offspring.
negative feedback – A control mechanism that reacts to a change in a condition (such as body temperature) by trying to bring the condition back to a normal level.
nematode worms – Small, unsegmented, usually microscopic, worms found in nearly every ecosystem on Earth.
nerve – Organ containing many neurones.
nervous system – An organ system that includes the brain and nerves, which carries information around an organism.
neurone – A cell that transmits electrical impulses in the nervous system.
neutralisation reaction – A reaction in which an acid reacts with a base to form water and a neutral salt.
neutron – Electrically neutral subatomic particle found in the nucleus of most atoms.
noble gases – Elements in group 0 of the periodic table.
non-infectious diseases – Disease not caused by an infecting agent (pathogen).
non-renewable energy – Energy source that cannot be replaced once it is used. It will run out one day.
nuclear energy – Potential energy that is stored in the nucleus of an atom.
nuclear fission – Nuclear reaction in which large nuclei break down to form small nuclei.
nuclear fusion – Nuclear reaction in which nuclei fuse together to form a bigger nucleus, releasing lots of energy. The energy released in a fusion reaction is greater than that from a fission reaction.
nuclear symbol – The chemical symbol for an atom showing the atomic number and the mass number.
nucleon number – The total number of protons and neutrons in the nucleus of an atom. Also known as the mass number.

ore – A rock from which metal can be extracted.
organelles – Small structures within cells. Each carries out a particular function.
organic fertiliser – Non-chemical fertiliser such as animal manure, plant compost or crushed animal bones.
osmosis – Movement of water from a region where there are many water molecules (high water potential) to a region of lower water potential through a partially permeable membrane, usually a cell membrane.
osteoporosis – A condition in which bones lose density and deteriorate. They are more likely to break.

pancreas – An organ that controls blood sugar concentration by producing insulin and glucagon. It also has another function, to produce digestive enzymes.
parallel – When electrical components are connected so that the same voltage is applied to each component.
particle accelerator – Device used to accelerate charged particles to very high speeds. Using particle accelerators it is possible for particle physicists to collide particles at high energies and through this to break down matter into smaller particles.
particle collision theory – The theory which states that for a reaction to happen between two substances, the particles of the substances must collide with enough force (or energy) for them to break up and re-bond to form new products.
pathogen – An organism that causes disease.
percentage yield (% yield) – A way of comparing the actual yield with the predicted yield for a reaction.
periodic table – A way of arranging elements in order of their atomic number to show their patterns and properties.
peripheral nervous system (PNS) – The nervous system outside of the brain and spinal cord.
pesticide – Substance that can kill an organism considered to be a pest.
phenotype – Visible characteristics of an individual.
photosynthesis – A set of chemical reactions in plants that allow them to produce their own food (glucose) using water and carbon dioxide and releasing oxygen as a waste product. The process is powered by light from the Sun.
photovoltaic cell – A solar cell that is used to convert solar energy to electricity.
physical property – Something we can observe about a substance that doesn't involve a chemical change; for example its appearance or melting point.
plagiarism – Using other people's work and presenting it as your own.
plate boundary – The location where two tectonic plates meet.
plate tectonics – A theory which is based on the idea that the Earth's crust is divided into separate tectonic plates. These tectonic plates float and move around on the liquid and mantle below the crust.
population – A group of organisms of the same species, living in the same ecosystem at the same time and able to interbreed.
potential energy – Stored energy that is either within the chemical structure of a substance (chemical potential energy) or is due to an object's position (gravitational potential energy).
predation – The killing and eating of one kind of organism by another kind of organism.
predator – An animal that survives by hunting and eating other organisms.
prey – Animals caught or hunted for food.
primary consumers – An alternative name for herbivores (animals that eat plants).
probability – The likelihood of something happening, often shown as a percentage chance. For example, there is a 50% chance that it will rain tomorrow.
producer – Organism that makes its own food, such as a plant using photosynthesis.
product – A substance which is formed at the end of a chemical reaction.
prokaryotes – Organisms that do not have their genetic material enclosed in a nucleus, nor do they have any other membrane-bound organelles. In most cases they are single-celled organisms, for example bacteria.
protoctists – One of the kingdoms of organisms. Most are single-celled, eukaryotic, but some are multicellular organisms. Includes algae, protozoa, plankton, seaweeds.
proton – Positively charged subatomic particle found in the nucleus of all atoms.
proton number – The number of protons in the nucleus of an atom. Also known as the atomic number.

radiation – The transfer of energy via electromagnetic waves.
radioactive – Releases radiation.
radioactive decay – When the nucleus of an unstable isotope changes by giving off radiation.

radiotherapy – The treatment of disease using X-rays, gamma rays or other ionising radiation.
random – When we can't say which particular nucleus is going to decay at a particular time.
reactants – The substances that react together in a chemical reaction.
receptors – Cells that receive a stimulus and convert it into an electrical impulse to be sent to the brain and/or spinal cord.
recessive – Allele that will only be expressed in the phenotype if there is no dominant allele of that gene present in the cells. Represented in genetic diagrams by a lower case letter.
red blood cell – Specialised cell containing haemoglobin (and has no nucleus or organelles and very little cytoplasm) that carries oxygen from the lungs to respiring cells.
red shift – The increase in wavelength (decrease in frequency) of electromagnetic radiation from distant, receding galaxies due to the expansion of the Universe.
reflex arc – Nerve pathway involving passage of impulses from sensory receptor, along sensory neurone to spinal cord and back, via motor neurone, to effector organ.
relative atomic mass – The average mass of an atom of an element compared to a standard mass.
renewable energy – Energy resource that will not run out, such as solar or wind energy.
resistance – How easy or difficult it is for an electric current to flow through something.
resistant – Bacteria that become resistant to antibiotics have the ability to survive treatment with antibiotics.
resistor – An electrical component that limits the flow of current.
respiration – The cellular process that all living organisms use to release energy from food to power their chemical reactions and activities.
reversible reaction – A chemical reaction that can work in both directions.
rickets – A childhood disease in which bones are weak; caused by a deficiency in vitamin D.
risk – A situation involving exposure to danger.
risk assessment – An investigation of the risks involved with a certain procedure and whether there is a safer alternative.
risk factors – Factors which greatly increase your chance of getting a certain disase but which do not always cause the disease, e.g. smoking tobacco, drinking too much alcohol.

Sankey diagrams – Block diagrams that show how energy is transformed, the width of the arrow representing the amount of energy.
scurvy – A condition affecting joints, blood vessels, gums and wound healing; also causes anaemia; caused by a deficiency of vitamin C.
secondary consumer – Animals that eat primary consumers. Also an alternative name for carnivores (animals that eat other animals).
semiconductor – A substance with moderately high electrical conductivity.
sensory neurone – A neurone that carries impulses from sensory receptors to the CNS (central nervous system – brain and spinal cord).
series – When electrical components are connected so that the same current flows through each component.
sickle cell disease – Genetic disorder caused by inheriting two copies of the recessive allele of the gene for making haemoglobin. The haemoglobin is faulty and this causes the red blood cells to be sickle shaped.
Solar System – The Sun and the planets and objects that orbit around it.
solubility – A measure of the amount of substance that will dissolve in a certain amount of solvent.
solution – A liquid formed by dissolving a solute in a solvent.
specialised (cell) – A cell that has special features that are required for its function. For example, red blood cells are specialised to carry oxygen.
species – Each different type of organism is called a species. The members of each species look and behave similarly, have similar genes and can reproduce together to produce fertile offspring (offspring that can also reproduce).
speed – A measure of how fast something is going.
spent fuel – The fuel that is left over after the nuclear fission reaction has finished.
state symbol – Abbreviation used to show the physical state of each substance in a reaction: (s), solid; (l), liquid; (g), gas; (aq) aqueous.
stimulus – A change to the external or internal environment of a cell/organism that causes a response (either physiological or behavioural). Receptors detect stimuli.
stomata (singular stoma) – Tiny pores in the lower surface of a leaf, which, when open, allow gases to diffuse into and out of the leaf.
structure – The way in which particles are arranged in chemical substances.
sustainable – Describes ways in which human beings can meet their needs without exhausting the world's resources or polluting the environment for future generations.
synapse – Point at which two neurones meet. There is a tiny gap between neurones at a synapse, which cannot transmit an electrical impulse. The information is carried across the synapse by special chemicals called neurotransmitters.

tectonic plate – A massive section of the Earth's surface that gradually moves around relative to other plates.
tertiary consumer – Top carnivores, animals that eat secondary consumers.
theory – An idea proposed as an explanation for a phenomenon. Scientists continue to gather evidence to support a theory but are always willing to modify the theory if they find new or conflicting evidence.
thermal decomposition – Reaction in which one substance breaks down when heated to form two or more new substances.
thermal energy – Energy transferred by heating.
thymine – One of the four bases in DNA. It forms a base pair with adenine.
transformation – When energy changes from one form to another.
transformer – A device that increases (step-up) or decreases (step-down) voltage.
trend – A change that occurs in a particular way, usually an increase or a decrease.
trophic level – One level of a food chain, such as producer, herbivore, carnivore.
tsunami – A huge wave caused by an earthquake or landslide on the sea bed.
turbine – A machine for producing power in which a wheel or rotor is made to revolve by a fast-movng flow of fluid.

vaccination – Procedure to inject a less harmful version of an infecting agent (microorganism) to make the body produce an immune response and make memory cells and so make the vaccinated organism immune to that pathogen.
vasoconstriction – Narrowing of the blood vessels (arterioles) near the surface of the skin.
vasodilation – Widening of the blood vessels (arterioles) near the surface of the skin.
vein – A vessel that carries deoxygenated blood returning to the heart.
virus – Extremely small infecting agent, not made of cells but made of nucleic acid and a protein coat, that infects cells so that those cells can make more copies of the virus. They cause diseases, e.g. flu, mumps, measles.
viscosity – How thick or runny a liquid is. Low viscosity is very runny, high viscosity is thick.
voltage – A measure of the amount of energy transferred by a current.
voltmeter – An instrument that measures voltage.
voluntary response – One under conscious control; involves thought.

watercourse – Any flowing body of water, such as rivers or streams, or pipes through which water may flow.
wavelength – The distance between a point on one wave and the same point on the next wave.
white blood cell – A type of cell that circulates in the blood and is an important part of the immune system. White blood cells are the body's main line of defence against disease. There are about six types of white blood cell.
word equation – A way of describing what happens in a chemical reaction. Word equations show you the names of the substances that react together and the new substances formed.

yield – The mass of product made in a reaction.

Index

The publisher would like to thank the following for their kind permission to reproduce their photographs:

(Key: b-bottom; c-centre; l-left; r-right; t-top)

Alamy Images: Adam Gault 115, Carpe Diem RF 161bl, Chad Ehlers 173t, David R. Frazier Photolibrary, Inc. 80r, Juice images 189, Kaith Morris 174, Monty Rakusen 145, Neil McAllister 185l, Radius Images 182, Alex Segre 68c, Steve Allen 177l, Steve Hamblin 89, Susan & Allan Parker 163; **CERN Geneva:** Copyright 2011 ATLAS Collaboration 154; **Corbis:** 109, Bettman 188l, David Frazier 106, dpa 34, Gary Braasch 110, Gyro Photography / Amanaimages 55r, Heritage Images 177r, Philip Evans / Visuals Unlimited 90, Dr. Richard Roscoe 44t; **DK Images:** 171b; **FLPA Images of Nature:** Reinhard Dirscherl 168r; **Getty Images:** Globo 42, Javier Larrea 184t, Lionel FLUSIN / Gamma-Rapho 37, Oxford Scientific / Photolibrary 113, Steve Gorton 103, Visuals Unlimited, Inc. / Tom Walker 119; **NASA:** 59, Hubble Heritage Team (STScI / AURA) / ESA 151r, J. Hester and A. Loll (Arizona State University) / ESA 148l, JPL- Caltech 126, SAO / CXC 148r, STScI 149, WMAP Science Team 155; **Nature Picture Library:** John Downer 160; **NCI:** 12; **Pearson Education Ltd:** Studio 8 10, 17, Gareth Boden 82, Jules Selmes 118, Martin Sookias 141t; **PhotoDisc:** StockTrek 134, 150t; **Photolibrary.com:** Michael Saint Maur Sheil / Photodisc 62; **Press Association Images:** Taro Konishi / AP 114; **Rex Features:** US Navy photo by John Gay 54br; **Science Photo Library Ltd:** Andrew Lambert Photography 92, Astier - Chru Lille 69l, Stephen Ausmus / US Department Of Agriculture 167, Martin Bond 65, Massimo Brega, The Lighthouse 180tl, DR JEREMY BURGESS 180b, Charles D. Winters 39t, 39b, 49, 98, Martyn F. Chillmaid 45, Cordelia Molloy 161tl, Dan Guravich 162r, David Mack 26, Claire Deprez / Reporters 50, Dr Keith Wheeler 169, 185r, Dr P. Marazzi 53, Drs A. Yazdani & D.J. Hornbaker 32, Vaughan Fleming 180c, Pr. G Gimenez-Martin 19, Mark Garlick 151l, Geoff Kidd 161br, Georgette Douwma 162l, Spencer Grant 21b, Peggy Greb / Us Department Of Agriculture 184b, Steve Gschmeissner 168l, Jan Hinsch 180bl, Mehau Kulyk 25, Leon J. Lebeau, custom Medical stock Photo 188b, Jeff Lepore 165, Matthew Oldfied 166, Tony Mcconnell 54bl, Astrid & Hans Frieder Michler 41r, Tom Myers / Agstockusa 172, Nigel Cattlin 161tr, David Parker 152, Pasieka 68r, Paul Rapson 95, 97, Power And Syred 168cl, Philippe Psaila 69br, Robert Brook 63b, Paul Silverman / Fundamental Photos 41c, Simon Fraser 192, Doug Sokell, Visuals Unlimited 180r, ST Mary's Hospital medical school 188t, Steve Percival 191l, Bjorn Svensson 168c, Victor De Schwanberg 116, Wayne Lawler 179, Mere Words 132; **Shutterstock.com:** a454 94l, AJP 129, Alxhar 186t, Anton Balazh 111, Hamiza Bakirci 168cr, Steve Ball 181t, Leonello Calvetti 29, CandyBox Images 22, Chris Jenner 63t, Danny Smythe 80l, David Kay 100, East 156, Elena Elisseeva 64t, 194, BW Folsom 41l, Guentermanaus 175, Holbox 133, Gabriela Insuratelu 24, itsmejust 69bl, Jason Steel 170, Sabine Kappel 43t, Sebastian Kaulitzki 21t, Kletr 178l, D. Kucharski & K. Kucharska 190, Emin Kuliyev 54t, Leah-Anne Thompson 31, Lee Prince 112, Daniel Loretto 183, Ilja Mašík 11, Mike Brake 57, Konstantin Mironov 150b, Mphot 48, mtr 181b, Stephen Mulcahey 61, Petr Nad 55l, Rob Byron 142, Rus Gri 138, Sakhorn 178, Alila Sao Mai 186b, Sergey Rusakov 68l, Studio 37 94, Subbotina Anna 171t, Wellford Tiller 52, Vasily Smirnov 83, Vladislav Gajic 173b, Volodymyr Krasyuk 80t, Warren Goldswain 67, wavebreakmedia ltd 191r, Yanas 80b, Yury Asotov 141b, Zoulou_55 66, zzoplanet 146; **STILL Pictures The Whole Earth Photo Library:** Biosphoto / Ribette Michel / BIOSphoto 176; **SuperStock:** fStop 96, Pixtal 44b, Prisma 69r, Radius Images 157, Robert Harding Picture Library 64b, Universal Images Group 43b

Cover images: *Front:* **Science Photo Library Ltd:** Mehau Kulyk